THE

NATURAL HISTORY

OF

BRITISH INSECTS;

EXPLAINING THEM

IN THEIR SEVERAL STATES,

WITH THE PERIODS OF THEIR TRANSFORMATIONS,
THEIR FOOD, ŒCONOMY, &c.

TOGETHER WITH THE

HISTORY OF SUCH MINUTE INSECTS

AS REQUIRE INVESTIGATION BY THE MICROSCOPE.

THE WHOLE ILLUSTRATED BY

COLOURED FIGURES,

DESIGNED AND EXECUTED FROM LIVING SPECIMENS.

By E. DONOVAN.

VOL. XII.

LONDON:

PRINTED FOR THE AUTHOR,
And for F. and C. RIVINGTON, Nº 62, ST. PAUL'S CHURCH-YARD,
MDCCCVII.

Printed by Law and Gilbert, St. John's Square, Clerkenwell.

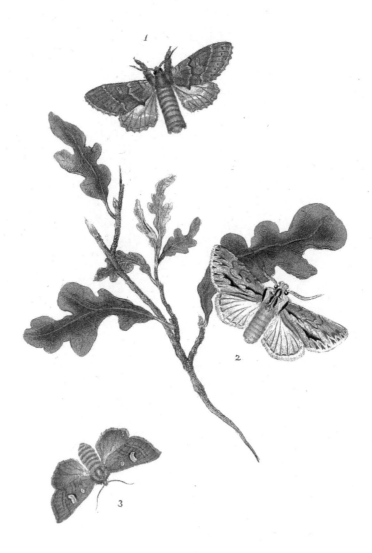

THE
NATURAL HISTORY
OF
BRITISH INSECTS.

PLATE CCCXCVII.

FIG. I.

PHALÆNA ZEBU.

ZEBU, PROMINENT.

LEPIDOPTERA.

GENERIC CHARACTER.

Antennæ gradually tapering from the bafe to the tip : wings in general deflected when at reft. Fly by night.

BOMBYX.

SPECIFIC CHARACTER

AND

SYNONYMS.

BOMBYX ZEBU. Wings deflected : back fingle toothed : thorax rufous : anterior wings pale rufous and fulvous varied, with two obsolete denticulate yellowifh bands.

VOL. XII.   B   LE CHAMEAU.

LE CHAMEAU. Chenille du Tremble, *var. e. f. Ernſt. Fig.*
267.
BOMBYX DROMEDARULUS. Small iron prominent. *Haworth,*
Lep. Brit. p. 101. *n.* 29.

Our Zebu Prominent, or, as the Aurelians call it, the ſmall Iron
Prominent, is extremely ſcarce. It was diſcovered in the larva ſtate
upon the oak : in the month of September it went into the ground
and became a pupa; the moth appeared in June following.

This inſect differs from the Fabrician Bombyx Dromedarius, or
what is termed with us the *Iron Prominent* in ſeveral reſpects, though
at the firſt view it ſeems to bear a very ſtrong reſemblance to that
ſpecies. We object to the trivial Engliſh name of ſmall Iron Prominent, becauſe it is only applicable, in a partial degree, to the inſect.
Our Zebu Prominent is certainly ſmaller than the inſect known
in this country by the name of Iron Prominent, but this is not invariably the caſe. We have ſeen the male of the Iron Prominent of a
ſize nearly, if not entirely, as diminutive as our ſpecimens of the Zebu
Prominent; and if we may rely on the accuracy of the figures of the
latter in the works of Ernſt, the Bombyx Dromedarius is not an inſect
of much ſuperior magnitude. We may truly infer from the figures
above mentioned, that the diminutive ſize of our Zebu is no criterion of the ſpecies.

A decided difference in point of colour, and in ſome other particulars, prevails between the two inſects. In Bombyx Dromedarius
the colour of the ſuperior wings is fuscous moſt delicately ſpeckled,
as it appears on cloſe inſpection, with grey; the ſpots of a deep
rusty-iron colour, and the denticulated bands acroſs the wings whitiſh,
distinct

PLATE CCCXCVII.   8

diſtinct, and well relieved with fufcous, and ferruginous. In our infect the general colour is pale rufous ſlightly tinged with fufcous in
the area of the wing, and varied towards the circumference with deep
fulvous : the denticulated bands acrofs are difpofed in a fimilar manner to thofe on the wings of Bombyx Dromedarius, but are of a yellowiſh inſtead of whitiſh colour, and nearly obſolete. So far as our
own obfervation extends there is a difference alfo in the pofterior
wings : in our Bombyx Zebu thofe wings are of a very pale fufcous
with only a fingle fainter band ; in Bombyx Dromedarius the wings
are paler ſtill ; it has likewife a fimilar band, but which is rather
more denticulated, and being bounded both above and below with a
duſky band, the wings appear of a lighter colour next the pofterior
margin, and in the diſk of the wing ; the latter part has alfo a fingle
ſhort transverse daſh of a duſky colour. A further difference is obfervable in the lower furface : the general tint in our B. Zebu is pale
ferruginous ; in B. Dromedarius greyiſh, with the lower pair whitiſh,
and in both, the bands confpicuous : the tip of the anterior pair in B.
Zebu teſtaceo-fufcous, in B. Dromedarius diſtinctly grey ; and the
central fpot in the lower wings of the latter fufcous with a white
fpeck in the center, but in B. Zebu plain teſtaceous without any
central mark.

In the " *Lepidoptera Britannica*," Mr. Haworth defcribes our
Bombyx Zebu as a fpecies perfectly new, under the name of Bombyx
Dromedarulus. The fpecimens from which his defcription is taken
were thofe in the cabinet of Mr. Drury, and which are now in our
poffeffion. This infect was probably new to the Aurelians of this
country, but certainly not fo to the continental entomologiſts, and thofe
ſhould affuredly have been confulted previouſly to its being defcribed as a
nondefcript infect. In the works of Ernſt which this ingenious writer
has overlooked, will be found a figure both of the upper and lower furface of the infect from a larger fpecimen than our own, and tolerably
expreffive ; and with a defcription of the infect equally fatisfactory.
It appears from thence that the figures are copied by Ernſt from a
female fpecimen in the noble collection of M. Gerning, of Frankfort,

B 2   which

which contains befides the male of the fame infect. Ernft is evidently in doubt whether to confider it as a variety of the Bombyx Dromedarius differing only from that infect in the gradations of colour, or as a diftinct fpecies; this he leaves for time and future refearches to determine. He obferves, however, and it is a ftrong argument in favour of its being diftinct, that the males in M. Gerning's collection, are of the fame colour as the female he reprefents, and the like circumftance is exemplified in our fpecimens *.

The variety fig. 1. of the B. Tritophus of Efper appears to be of the fame fpecies as our B. Zebu, but of this we cannot fpeak with confidence. Schneider certainly notices it. Fabricius probably confidered it as a variety of B. Dromedarius.

F I G. II.

PHALÆNA CASSINIA.

TRILINEATED MOTH.

SPECIFIC CHARACTER

AND

SYNONYMS.

Wings deflected, grey with abbreviated fcattered black lines: thorax with a black line each fide, and in the middle.

---

* " M. Gerning qui poffédé dans fa collection l'individu femelle dont nous donnons le portrait en deffus et en deffous, *fig.* 267. *e, f,* le croit une variété de cette efpèce, malgré la différence que l'on y remarque dans les nuances. Cependant comme il a des mâles de la même couleur, il n'ofe affurer que ce ne foit point une efpèce différente. Le temps feul et les referches peuvent nos en rendre certain." *Ernft.*

BOMBYX

PLATE CCCXCVII. 5

BOMBYX CASSINIA: alis deflexis grifeis: lineolis abbreviatis nigris
fparfis. *Fabr. Ent. Syft. T. 3. p. I.* 460. *n.* 164.
B. CASSINIUS. SPRAWLER. *Haw. Lep. Brit. p.* 106. *n.* 40?

Fabricius defcribes his Bombyx Caffinia as a native of Auftria
from the cabinet of Schieffermyller. It is found on the Lime. If this
be of the fame fpecies as the B. Caffinius above quoted, it is alfo found
in the larva ftate on the oak, and appears in the winged ftate in September.

This infect, confidered as a Britifh fpecies, is almoft equally as fcarce
as the preceding, Bombyx Zebu.

FIG. III.

NOCTUA AURICULA.

GOLDEN EAR MOTH.

SPECIFIC CHARACTER

AND

SYNONYMS.

NOCTUA AURICULA. Anterior wings fub-ferruginous with a fmall
fulvous fpot, and in the middle a larger ear-fhaped
yellow fpot enclofing a lunar ring.
L'ECLATANTE. *Ernft.* II. *part. v.* 6. *n.* 394.

This

This infect has been erroneoufly confidered by Efper and others as the Phalæna nictitans of Linnæus, an infect which it pretty much refembles, but from which it differs fpecifically. This circumftance is mentioned particularly by Ernft, who defcribes and figures both the Linnæan fpecies, and the infect miftaken for it. The fpecimens he delineates are in the cabinet of M. Gerning of Frankfort. Ernft defcribes our infect as a fcarce fpecies in Germany: in England we believe it is very rare; the only fpecimen we poffefs is in the cabinet of Mr. Drury.

PLATE

PLATE CCCXCVIII.

TENTHREDO FASCIATA.

BANDED SAW-FLY.

HYMENOPTERA.

GENERIC CHARACTER.

Mouth with a horny mandible, curved, and toothed within ; jaw straight and obtufe at the tips : lip cylindrical and three cleft : feelers four unequal and filiform : wings flat and tumid : fting compofed of two ferrated laminæ and fcarcely difclofed.

SPECIFIC CHARACTER

AND

SYNONYMS.

Deep black : antennæ black : upper wings with a fufcous band.

Tenthredo Fasciata : atra, antennis nigris, alis primoribus fafcia fufca. *Fabr. Sp, Inf.* 1. *p.* 407. *N.* 8.—*Gmel. Syft. Nat.* 2655. 7.

Tenthredo antennis clavatis nigris, abdomine glabro atro, alis fuperioribus fafcia fufca. *Linn. Syft. Nat.* 12. 2. *p.* 921, *n.* 7.—*Fn. Suec.* 1538.

This is a rare infect in England. Linnæus defcribes it as a native of Sweden, Panzer as a German fpecies, and by other writers it is mentioned as a general inhabitant of Europe.

Fabricius

Fabricius fpeaks of a fmall white band at the bafe of the firft abdominal fegment of this fpecies: in our Britifh fpecimen, the whole of
this fegment is of a pale or whitifh colour inftead of the bafe only, and
it appears from the figure given by Panzer, *Fn. Germ.* that the fame
circumftance is obfervable in the individual he has delineated.

The larva, and metamorphofes of this fpecies of Tenthredo has not
hitherto been afcertained by any writer.

PLATE

PLATE CCCXCIX.

FIG. I. I.

APIS DRURIELLA.

DRURY'S BEE.

HYMENOPTERA.

GENERIC CHARCTER.

Mouth horny: jaw and lip membranaceous at the tip: tongue infle&ted: feelers four, unequal, filiform: antennæ fhort, and filiform; thofe of the female fomewhat clavated: fting of the females and neuters pungent, and concealed within the abdomen.

SPECIFIC CHARACTER

AND

SYNONYMS.

Black with cinereous down: antennæ fulvous beneath: laft joints of the abdomen mucronated on each fide.

Apis Druriella: nigra, hirfuto cinerafcens; antennis fubtus fulvis; abdomine fegmentis pofticis utrinque mucronatis. *Kirby. Ap. Angl. v.* 2. *p.* 285. *n.* 62.

C

This

This very uncommon ſpecies of Apis is not mentioned by any entomological writer, except Mr. Kirby, whoſe "Monographia Apum Angliæ" affords a minute deſcription of it. Mr. Kirby deſcribed it from a ſpecimen in the cabinet of the late Mr. Drury, and aſſigned it the name of Druriella, in compliment to that zealous collector and writer; and we are perfectly diſpoſed, for the ſame reaſon, to adopt it. It is almoſt needleſs to add, that being in poſſeſſion of the Engliſh entomological cabinet of the late Mr. Drury, our figures are delineated from the individual inſect Mr. Kirby deſcribed.

The ſmalleſt inſect at fig. 1. I. repreſents Apis Druriella in its natural ſize, that above exhibits its magnified appearance.

FIG. II. II.

APIS VARIEGATA.

VARIEGATED BEE.

SPECIFIC CHARACTER

AND

SYNONYMS.

Thorax and abdomen variegated with white: legs ferruginous.

APIS VARIEGATA: thorace abdomineque albo variegatis, pedibus
ferrugineis. *Linn. Fn. Suec.* 1699.

NOMADA VARIEGATA: *Fabr. Ent. Syſt. T.* 2. *p.* 347. *n.* 5.

APIS VARIEGATA: nigra; trunco, abdomineque, albido variegatis;
pedibus ferrugineis. *Kirby Apium Angl. p.* 222.
n. 86.
Panzer. Fn. Inſ. Germ. Init. n. 61. *tab.* 20.
Forſt. Cat. Brit. Inſ. n. 1033.
Apis. n. 26.

Apis muſcaria, *Chriſtii. Hymenop. p.* 195. *tab.* 17. *fig.* 5.

Apis

PLATE CCCXCIX 11

Apis variegata, though an infect of a fmall fize, is interefting for its rarity and elegance. This pretty fpecies was firft introduced to obfervation, as a Britifh Infect, by Dr. Forfter, in his " *Novæ Species Infectorum Centuria prima.*" Since his time, it feems to have been mentioned only by Mr. Kirby, who, in his work entitled *Monographia Apum Angliæ,* obferves, that he firft faw this bee in the cabinet of Mr. Drury, and afterwards found it, but by no means common, in the autumn of two fucceeding years, 1797 and 1798, flying about funny banks; it is remarkable, that after the time laft mentioned he never met with it.

This infect is liable to vary a little in colour in different individuals; and it is further obfervable, that the Englifh fpecimens are fmaller than thofe found in other parts of Europe.

C 2

PLATE

PLATE CCCC.

LUCANUS INERMIS.

SHORT-HORNED STAG BEETLE.

COLEOPTERA.

GENERIC CHARACTER.

Antennæ clavated, the club compreffed and divided into pectinated leaves: jaws projecting and dentated: two palpigerous tufts under the lip: body oblong: anterior fhanks dentated.

SPECIFIC CHARACTER

AND

SYNONYMS.

Convex, brown: jaws fhort, with raifed lateral teeth.

LUCANUS INERMIS: convexus brunneus, maxillis brevibus, dente laterali elevato. *Marfh. Ent. Brit. T. I. p.* 48. *n.* 2.

LUCANUS CERVUS. *Linn. Faun. Suec.* 405.

Lucanus Cervus. *Var. β. Gmel.* 1588. *I. Fabr. Ent. Syft.* 1. *p.* 2. 236. 2. *Var. β.*

Lucanus Dorcas. *Harr.* 5. 2.

Platycerus. La Grande Biche. *Geoffr. I.* 62. 2.

The

The Short-horned Stag-beetle is confidered by almoſt every writer either as the female, or as a variety of the Linnæan Lucanus Cervus. This opinion is controverted by Mr. Marſham, who informs us in his *Entomologia Britannica*, that he has taken the two ſexes of the Cervus Lucanus together, in a ſtate that can admit of no doubt, that they were really the two ſexes of that ſpecies; or, at leaſt, that there are both males and females of the long-horned kind. At the ſame time the evidence of Geoffroy is adduced to prove ſtill further, that there are two diſtinĉt ſexes of the Short-horned Stag Beetle, the infeĉt hitherto believed by many to be the true female of Cervus Lucanus.

The accuracy of the obſervations, upon which the aſſertions of the above mentioned authors are founded, admits of little diſpute: that the infeĉts in queſtion have been found in the ſituation before intimated we muſt readily believe; but as this might happen whether they were in reality of the ſame ſpecies or not; and as the opinion generally ſupported by authors of the firſt reſpeĉtability is in favour of the Short-horned Stag Beetle being the female of the other kind, it is proper we ſhould ſay a few words further reſpeĉting them.

Roeſel, in his deſcription of the Cervus Lucanus, expreſsly tells us, he has found the male of that infeĉt and the Short-horned Stag together, and concludes that they are of the ſame ſpecies. To this may be added the authority of Linnæus, who caught them in the ſame ſituation, and naturally inferred, for the ſame reaſon, that they were the two ſexes of an individual ſpecies. Neither are thoſe the only writers, who mention the ſame circumſtance; and beſides thoſe, the faĉt is ſufficiently well known to many praĉtical colleĉtors, who have obſerved them in the ſame ſtate, without in the leaſt ſuſpeĉting that they were in reality the male and female of two diſtinĉt, though very analogous ſpecies.

When

PLATE CCCC. 15

When two infects, however diffimilar in appearance, occur together in this ftate, it is a natural conclufion, that they are the two fexes of the fame fpecies. This is pretty generally, but not invariably the fact. It requires only a very curfory attention, for example, to the genera of Cicada and Coccinella to prove, that the moft promifcuous intercourfe prevails between the two fexes of the greater number of fpecies in thofe extenfive genera ; and that the varieties arifing from this intercoufe of the fexes are the fource of inexplicable confufion to the entomologift; a fpurious brood being by that means intro-duced, that cannot eafily be reduced to either of the parent species. The fame applies, though certainly with a lefs degree of latitude, to fome larger infects, efpecially in the Coleoptera tribe. We muft allow, that, though it is almoft a conclufive evidence, when we find infects of the two fexes coupled together, that they are of the fame fpecies; but it does not follow, as a matter of certainty, that they are fuch : the conclufion is fpecious, and in general correct, but we cannot always depend on it. Even fo it appears with regard to the Long-horned and Short-horned Stag Beetles : when we find, as is not unfrequently the cafe, thofe two infects connected together, we conclude, they are the true male and female of the fame fpecies ; and probably without further ex-amination affent to the popular notion, that the horned kind is the male, and the hornlefs fort the female, whereas perhaps the very reverfe might with a flight attention be fometimes difcovered ; we might detect the horned female with the hornlefs male. It is a little remarkable we muft indeed confefs, in admitting that there are males and females of both kinds, that thofe rovers fhould fo rarely occur in connection with the infects, which nature has ordained as their refpective mates.

It has been previoufly remarked, that we may reft affured at leaft, that there are males as well as females both of the Short and the Long-horned Stag Beetles. Geoffroy is believed to have been the firft writer, who difcovered the error of confounding the former with the female of the latter : he defcribes the Short-horned kind under the

the name of La Grande Biche, and tells us he has frequently seen both sexes of this insect coupled together*. He observes, however, that he has never seen the two sexes of the long-horned kind in the same state, a circumstance that in our mind rather weakens than confirms the strength of his former remark. Neither does Geoffroy take notice of any difference of appearance between the male and female of his Grand Biche, which leaves us in further doubt. If, notwithstanding, we can rely upon this writer, one disputed fact may be collected from the result of his remarks, namely, that there are both males and females of the Short-horned Stag Beetle.

This point attained, our attention is next directed to discover the true female of Cervus Lucanus; and this, if we are not mistaken, has been noticed only by Mr. Marsham. This gentleman, as it appears from the *Entomologia Britannica*, was so fortunate as to take the two sexes together, some years ago. The female is described as being in no respect different from the male, except in size, which is smaller; the horns are as large in proportion, they are beset with about seven nearly equal and approximate teeth, and have not one larger and remote from the rest. The discovery of the female of the Lucanus Cervus was a desirable circumstance; not merely as being the means of ascertaining the history of that insect, but as adding a further confirmation to the observations of Geoffroy, with regard to the Short-horned Stag, which they seem to us to require†. We have no doubt, that Geoffroy discovered both sexes of the Grande Biche, but it would have been more satisfactory had he been acquainted with the true female of the Cervus Lucanus. We need

---

* " Cette animal (La Grande Biche) resemble beaucoup au précédent; quelques personnes même ont cru qu'il n'en différoit que par la sexe, prenant celui-ci pour la femelle, et le cerf-volant pour la mâle : mais quoiquils se resemblent beaucoup pour la forme, la grandeur, et la couleur, il est prouférent pas seulement par le sexe, ayant rencontré plusieurs fois des biches accouplées ensemble, et jamais avec des cerfs volans.

† " Et nos etiam Geoffroyii sententiam comprobamus, cornutos enim copulâ conjunctus cepimus." *Marsh. Ent. Brit. T. I. p.* 48. *n.* 2.

only

PLATE CCCC.   17

only add, that no doubt can remain as to the two Long-horned Stag Beetles, taken by Mr. Marſham, being a male and female, as they were diſſected by Mr. Leman to determine the fact with precifion.

The Short-horned Stag Beetle has the fame haunts as the Cervus Lucanus, being found chiefly in the trunks of old or rotten trees, and is not uncommon.

PLATE CCCCI.

FIG. I.

MUSCA PYRASTRI.

DIPTERA.

GENERIC CHARACTER.

Mouth with a foft exferted flefhy probofcis, and two unequal lips: fucker befet with briftles: feelers fhort and two in number, or fometimes none: antennæ ufually fhort.

✳ Antennæ a naked briftle.

SPECIFIC CHARACTER

AND

SYNONYMS.

Almoft naked, black: thorax immaculate: abdomen with three pair of recurved whitifh lunules.

MUSCA PYRASTRI: nudiufcula, nigra, thorace immaculato, abdomine bis tribus lunulis albis recurvatis. *Linn. Fn. Suec.* 1817.
Scop. Ent. Carn. 931.
Gmel. Linn. Syft. Nat. 2875. *fp.* 51.

SYRPHUS PYRASTRI. *Fabr. Spec. Inf.* 2. *p.* 432. *n.* 58.—*Mant. Inf.* 2. *p.* 340. *n.* 67.—*Ent. Syft. T.* 4. 305. *n.* 102.

D 2

MUSCA

Musca Rosæ. *De Geer. Inf.* 6. *p.* 108., *n.* 5. *t.* 6. *fig.* 18.
Mufca thorace nigro-viridi, abdomine atro ovato, tribus paribus
      lunularum albicantium.—La mouche à fix taches
      blanches en croiffant fur le ventre. *Geoffr. Inf.* 2.
      517. *n.* 46.
      Frifch. Inf. 11. *t.* 22. *f.* 1.
      Reaum. Inf. 3. *t.* 31. *f.* 9.

The larva of this Infect feeds on the fpecies of aphis that infefts
the common pear; it is of a fine green colour, with a fingle yellowifh
white dorfal line, extending the whole length, from the head to the
extremity of the tail. The winged infect occurs in gardens.

FIG. II.

MUSCA CAEMETERIORUM.

SPECIFIC CHARACTER

AND

SYNONYMS.

Braffy black: abdomen depreffed, black, and fhining: wings
blackifh.

Musca Caemeteriorum: nigro aeneus abdomine depreffo atro
      nitido, alis nigricantibus. *Linn. Fn. Suec.* 1842.
      Syft. Nat. 2. 992. 82.
Syrphus Caemeteriorum: *Fabr. Ent. Syft. T.* 4. *p.* 303. *n.* 94.

Inhabits Sweden, and other parts of Europe, as well as England.

                             FIG.

PLATE CCCCI.

21

FIG. III.

MUSCA RIBESII.

RED CURRANT-LOUSE FLY.

SPECIFIC CHARACTER

AND

SYNONYMS.

Almoſt naked: thorax immaculate: abdomen with four yellow belts, the firſt interrupted.

MUSCA RIBESCII: nigra nudiuſcula, thorace immaculato, abdomine cingulis quatuor flavis: primo interrupto. *Linn. Fn. Suec.* 1817.—*Gmel. Linn. Syſt. Nat.* 2875. *n.* 50.

————

Feeds on the aphides ribis, or plant-louſe, that infeſts the red currant.

This inſect agrees with the Linnæan ſpecific deſcription of Ribeſcii; yet we muſt obſerve, that the extreme ſegment of the abdomen being yellow, ſeems to form a fifth, or additional yellow band, to the four, deſcribed by that author and by Fabricius.

PLATE

PLATE CCCCII.

TENTHREDO SERICEA.

SILKY SAW-FLY.

HYMENOPTERA.

GENERIC CHARACTER

Mouth with a horny mandible, curved and toothed within: jaw ftraight and obtufe at the tips: lips cylindrical and three cleft: feelers four unequal and filiform: wings flat and tumid: fting compofed of two ferrated laminæ, and fcarcely difclofed.

SPECIFIC CHARACTER

AND

SYNONYMS.

Antennæ clavated, reddifh, or black; abdomen green, or dufky, and bronzed.

Tenthredo Sericea: antennis clavatis luteis, thorace atro: abdomine aeneo. *Linn. Syft. Nat.* 2. 921. 8.— *Schaeff. Elem. Tab.* 51. β Tenthredo nitens antennis clavatis luteis, abdomine viridi cœrule-fcente nitente. *Linn. Syft. Nat.* 2. 922. 10.— *Fn. Suec.* 1532.—*Sulz. Inf. tab.* 18. *fig.* 109.

═══════════

The two fexes of Tenthredo Sericea differ fo greatly from each other, that fome authors have miftaken them for diftinct fpecies.

Linnæus

Linnæus deſcribes the male as a variety of the female. The male inſect, which we have repreſented, is of a beautiful ſilky greeniſh colour, gloſſed with a braſſy luſtre, the antennæ clavated and brown: legs yellow, except at the baſe of the thighs, which are black, and the five laſt joints of the abdomen marked in the center with a broad ſtripe of dark, or velvetty black. The female has the colours throughout more obſcure, the antennæ are duſky, approaching black; the thorax dark, and the abdomen braſſy, but tinged with a ſombrous hue inſtead of green, and the wings are rather darker than in the male.

This elegant ſpecies is inſerted among our Britiſh Inſects upon the authority of two ſpecimens, which we are credibly informed were found in England. Fabricius deſcribes it as the offspring of a ſhort green-coloured larva, which is marked with two yellow lines, and has a cinereous head, with a reddiſh brown band. It feeds on the alder.

PLATE

403

1

2

2
*

PLATE CCCCIII.

FIG. I.

APIS IRICOLOR.

IRICOLOR BEE.

HYMENOPTERA.

GENERIC CHARACTER.

Mouth horny: jaw and lip membranaceous at the tip: tongue inflected: feelers four, unequal, filiform: antennæ fhort, and filiform; thofe of the female fomewhat clavated: fting of the females and neuters pungent, and concealed within the abdomen.

SPECIFIC CHARACTER

AND

SYNONYMS.

Violaceous, above glabrous: wings blackifh.

APIS IRICOLOR: violacea, fupra glabra; alis nigricantibus. *Kirby, Ap. Angl. T. 2. p. 310. n. 72.*
Drury Inf. Vol. I. p. 108. tab. 45. fig. 3.
APIS VIRENS: *Chriftii Hymenop. p. 123. tab. 6, fig. 2.*

━━━━━━━━

Apis Iricolor is, a large and beautiful fpecies, but which, we are almoft perfuaded, has been introduced into the Britifh Catalogue without fufficient reafon. Mr. Kirby found it in Dr. Latham's

VOL. XII.   E   cabinet

cabinet among his Englifh apes; but Dr. Latham did not recollect where it was taken, or upon what authority he confiders it as Britifh. Notwithftanding this, Mr. Kirby has inferted the fpecies in his " Monographia Apum Anglia", and, in compliance with this authority, we have ventured to introduce it into the' prefent Work; conceiving, that after this explicit avowal of our only motive for enumerating it among the Britifh Infects, we fhall not be deemed entirely refponfible for the accuracy of our information. Apis iricolor is well-known as a native of the Weft Indies. Mr. Drury's work on Exotic Infects contains the figure of a fpecimen he received from the ifland of Jamaica. It very much refembles the Linnæan apis violacea, but has the body violaceous inftead of black, and the wings blackifh inftead of violet. We fhould obferve, that the wings in our fpecimen of Apis Iricolor does not appear to be fo dark or blackifh in colour as in the individual figured and defcribed by Drury; they are dufky, rather inclining to brown, and flightly gloffed with green: the thorax fine blue, and very glabrous; abdomen inclining more to greenifh, and the fegments edged at the bafe with fine reddifh purple.

FIG. II. II.

APIS BANKSIANA.

BANKSIAN BEE.

SPECIFIC CHARACTER

AND

SYNONYMS.

Deep black, fhining, glabrous; claws rufous.

APIS BANKSIANA: atra, nitida, glabriufcula; digitis rufis. *Kirby Ap. Angl. T. 2. p.* 179. *n. 3.*

Very

PLATE CCCCIII. 27

Very fimilar to the Apis Linnæella of Kirby, but twice its fize: Apis Linnæella is likewife diftinguifhed by having the extreme half of the antennae rufous; whereas in the Apis Bankfiana, the whole of the antennæ is black. This new fpecies is named after Sir Jofeph Banks, Bart. It is a rare infect.

The fmalleft figure denotes the natural fize.

E 2   PLATE

PLATE CCCCIV.

FIG. I.

SCARABÆUS FIMETARIUS.

COLEOPTERA.

GENERIC CHARACTER.

Antennæ clavated, the club fiffile: fhanks of the anterior legs generally dentated.

SPECIFIC CHARACTER

AND

SYNONYMS.

Head tuberculated: wing-cafes red: body black.

Scararæus Fimetarius: capite tuberculato, elytris rubris, corpore nigro. *Marfh. Ent. Brit. T.* 1. *p.* 10. *n.* 7.

Scarabæus Fimetarius: ater, capite tuberculato, elytris rubris. *Linn. Fn. Suec.* 385.—*Syft. Nat.* 548. 32.— *Fabr. fp. Inf.* 1. *p.* 16. *n.* 64. *Ent. Syft.* 1. 27. 84.

Scarabæus pilularius nonus. *Raj. Inf. p.* 106. *n.* 9.

Le Scarabé Bedeau: *Geoffr. Inf.* 1. *p.* 81. *n.* 18.

Very common in the dung of cattle.

In

The head of this infect is black: the fhield of the head fubrotund: thorax punctured and black, with a large anterior teftaceous fpot on each fide. The colour of the wing-cafes vary from rufous brown to reddifh: the legs are black, except the tarfi, which are rufous.

FIG. II.

SCARABÆUS SORDIDUS.

SPECIFIC CHARACTER

AND

SYNONYMS.

Head tuberculated: thorace black: margin pale with a black dot on each fide: wing-cafes teftaceous.

SCARABÆUS SORDIDUS: capite tuberculato, thorace nigro: margine pallido: puncto nigro, elytris-teftaceis. *Marfh. Ent. Syft. T.* 1. *p.* 10. *n.* 6.

SCARABÆUS SORDIDUS: capite tuberculato, thorace nigro: margine pallido, puncto nigro, elytris grifeis. *Fabr. Spec. Inf. I.* 17. 68.—*Syft. Ent.* 16. 55.—*Ent. Syft. I.* 29. 90.—*Schaeff. Icon. t.* 74. *f.* 3.

This infect is the fame fize as the preceding, the figure in our plate being magnified; and, like that fpecies, it is found in horfe-dung, but lefs frequently.

The antennæ are pale: head pale with three tuberculations: thorax punctured, black, with an entire pale or reddifh border, which is

broadeft

PLATE CCCCIV.   31

broadeſt at the ſides, and a ſingle lateral dot of black: wing caſes ſordid teſtaceous, and marked in general with two black dots each, but which in ſome ſpecimens are ſcarcely viſible.

FIG. III.

SCARABÆUS SORDIDUS. *Var.*

Among other varieties of Scarabæus sordidus, we ſometimes obſerve it without any of thoſe black ſpots, which appear ſo conſpicuous on the wing-caſes of the inſect delineated at No. 2. An enlarged figure of this immaculate variety is repreſented at fig. 3.

FIG. IV.

SCARABÆUS COPRINUS.

SPECIFIC CHARACTER

AND

SYNONYMS.

Deep black, wing-caſes teſtaceous with dotted ſtriæ, and black ſuture.

SCARABÆUS COPRINUS: ater, elytris punctato-ſtriatis teſtaceis: ſutura nigra. *Marſh. Ent. Syſt. T.* 1. *p.* 12. *n.* 11.
SCARABÆUS SORDENS. *Gmel.* 1546. 413?

Found in dung. This is a ſcarce ſpecies, and has not been noticed by Fabricius.

PLATE

PLATE CCCCV.

PAPILIO PILOSELLÆ.

LARGE HEATH, OR GATE-KEEPER BUTTERFLY.

LEPIDOPTERA.

GENERIC CHARACTER.

Antennæ terminated in a club: wings erect when at reft: fly by day.

SPECIFIC CHARACTER

AND

SYNONYMS.

Wings indented, fufcous, with fulvous difk: anterior pair with a bipupillated black fpot near the tip; and fnowy dots on the pofterior wings beneath.

PAPILIO PILOSELLÆ: alis dentatis fufcis: difco fulvo, anticis utrinque ocello nigro: pupilla gemina, pofticis fubtus punctis ocellaribus niveis. *Linn. Syft. Mant. I.* 537.

PAPILIO PILOSELLÆ: *Fabr. Ent. Syft.* 3. 240. 748.—*Syft. Ent.* 497. 233.

Papilio Tithonus. *Fabr. Spec. Inf.* 80. 355.

PAPILIO PILOSELLÆ: *Gmel. Linn. Syft. Nat.* 2300. *n.* 552.

PAPILILIO PILOSELLÆ, LARGE HEATH. *Haworth Lep. Brit. p.* 24. *n.* 28.

VOL. XII.   F   PAPILIO

PAPILIO HERSE. *Wien Schmetterl. p.* 320. *n.* 24.
PAPILIO TITHONUS: *Lewin. Pap.* 22.

———

    This is one of the moſt abundant of the European Papiliones. In England it literally ſwarms about the hedges in the month of July, when it appears in the winged ſtate. The larva is greeniſh, with a white line, and brown head: it is found in this ſtate in June.— The male of this butterfly is rather ſmaller than the female, and is diſtinguiſhed further by having an oblique duſky band acroſs the fulvous diſk of the anterior wings.

PLATE

PLATE CCCCVI.

PHALÆNA CHI.

CHI MOTH.

LEPIDOPTERA.

Noctua.

GENERIC CHARACTER.

Antennæ gradually tapering from the bafe to the tip: wings in general deflected when at reft. Fly by night.

SPECIFIC CHARACTER

AND

SYNONYMS.

Wings hoary grey: anterior pair marked with a black χ.

PHALÆNA CHI: alis canis: fuperioribus χ nigro notatis. *Linn. Fn. Suec.* 1180.
NOCTUA CHI: *Fabr. Spec. Inf.* 2. *p.* 236. *n.* 130.—*Mant. Inf.* 2. *p.* 174. *n.* 258.—*Ent. Syft. T.* 3. *p.* 2. *p.* 107. *n.* 321.

Phalena feticornis fpirilingius, alis deflexis, fuperioribus cinereo fufcoque nebulofis, lineis undulatis et omicro nigris, inferioribus cinereis. L'OMICRON NÉBULEUX *Geoffr. Inf.* 2. *p.* 156. *n.* 93.
Albin Inf. t. 83. *f.* C. D.
Roe. Inf. 1. *Phal.* 2. *t.* 13.

F 2   During

During one of our fummer excurfions through the northern parts of the principality of Wales, we difcovered by accident a folitary fpecimen of Phalæna Chi, refting among the lichens that encruft the fide of that venerable memorial of Druidical antiquity, the larger *Cromlech*, at *Plas Newydd* in the ifle of Anglefea. We obferved it in a perfectly quiefcent ftate, apparently juft as it had emerged from the pupa in the open day time. This was on the 17th day of Auguft: we conceive it requifite to mention this precifely, as the time of its appearance in the winged ftate has been ftated by one writer to be July inftead of Auguft, and that on the authority of the individual fpecimen above mentioned *.

An erroneous reference among the Fabrician fynonyms of Phalæna Chi, to the work of Albin (Plate 83), had long fince induced the Englifh collectors to confider the Phalæna Chi as a native of this country; but under this impreffion they miftook a very different infect for the Chi of Linnæus. Indeed the infect figured in Plate 83 of the work of Albin is no other than the common Sycamore Moth (*Phal. Aceris* of *Brit. Inf.*). It is therefore clear, this incautious reference has been productive of confiderable mifunderftanding, and that the Phalæna Chi, though really Britifh, was not introduced with propriety as fuch, till we difcovered it in Anglefea. We fhould however add, that among a number of infects found in Yorkfhire, and communicated for our infpection by Mr. Rippon, of York, we found an infect very analogous, and apparently a variety of the fame fpecies, though of a darker colour than our Cambrian fpecimen, which in this refpect accords extremely well with the Linnæan expreffion, " *alis canis*." It fhould be remarked, that Geoffroy, who defcribes Phalæna Chi, as an infect by no means uncommon in France, tells us, it varies much in fize, and fpeaks of varieties that are reddifh, and others

---

* Noctua *Chiina. Jul. Septis. Haworth* Prod. Lep. Brit.

blackifh,

PLATE CCCCVI.   37

blackifh *, though they agree in the markings; and he further obferves, that the female is of a more cinereous colour than the male,
but has the fame kind of fpots on the wings.

Our figure of the larva is taken from a well-preferved fpecimen
lately obtained by us from Germany as the larva of this fpecies,
and which agrees fo exaƈtly with the Linnæan defcription, and the
figure in the plate of Roefel, that we cannot hefitate to admit it as
fuch. The pupa delineated in our plate is alfo from the fame fource.
The food of the Phalæna Chi is the columbine and the thiftle.

" Il y a des variétes de cette Phalêne, qui font rougeâtres et d'autres noirâtres:
mais toutes ont les deux taches ronde et quarrée fur les aîlés.

PLATE

Geoffroy ſpeaks of this ſpecies of Dragon-fly as a rare inſect in France, and we believe it to be very uncommon in England. The only ſpecimen we have ſeen is one we poſſeſs in the cabinet of Mr. Drury, but from whence he obtained it we cannot aſcertain. Ray appears to have deſcribed this inſect as a native of Britain *.

This ſpecies bears ſome reſemblance to the female of Libellula Depreſſa †, in ſize and general appearance ; but on the ſlighteſt in-ſpection will be found entirely diſtinct. The abdomen is leſs depreſſed on the back and narrower, and being beſides ſomewhat compreſſed at the ſides, aſſumes a much more linear form in the middle than Libellula Depreſſa. The wings are totally different, and afford an excellent characteriſtic by which the ſpecies may be diſtinguiſhed. Each of thoſe wings are marked on the anterior or coſtal rib, with two ſmall brown ſpots, the one of which is ſituated near the apex, as in moſt of the tribe; the other about the middle on the anterior part of each wing : all the wings are yellowiſh at the baſe, and in the poſterior pair, that part is diſtinguiſhed further by having a large brown ſpot contiguous to the body.

---

* *Ray Inſ. p. 49. n. 3.*
† *Brit. Inſ. Vol. I. pl. 24.*

PLATE

PLATE CCCCVIII.

FIG. I. I.

APIS CORNIGERA.

HYMENOPTERA.

GENERIC CHARACTER.

Mouth horny: jaw and lip membranaceous at the tip: tongue inflected: feelers four, unequal, filiform: antennæ fhort, and filiform; thofe of the female fomewhat clavated: fting of the females and neuters pungent, and concealed within the abdomen.

SPECIFIC CHARACTER

AND

SYNONYMS.

Deep black: lip fomewhat cornuted: firft abdominal fegment with yellow fpots on the back: beneath ferruginous: abdomen variegated with yellow bands.

APIS CORNIGERA: atra, labio fubcornuto; abdominis ventre bafi ferrugineo, tergo maculis, faciifque flavis variegato. *Kirby Ap. Angl. T. 2. p.* 190. *n.* 11.

In the Apum Angliæ above quoted it is obferved, that when this infect is alive, it emits a very agreeable fcent. We have never

VOL. XII. G taken

taken this individual fpecies alive ourfelves; but we have obferved a fimilar odour emitted by other infects of the fame family.—Apis Cornigera is an infect liable to confiderable variations; we poffefs feveral infects in our own cabinet, which appear to be varieties, as they poffefs the fame characteriftic fpecific marks in general, though they differ in fome other particulars. It is an elegant and interefting infect.

FIG. II. II.

APIS JACOBÆÆ.

SPECIFIC CHARACTER

AND

SYNONYMS.

Deep black: legs ferruginous: abdomen with fix yellow bands, the three firft interrupted: antennæ above black.

APIS JACOBÆÆ: atra; pedibus ferrugineis; fcutelli punctis, abdominifque maculis fafciifque, flavis; antennis fupra nigris. *Kirby Ap. Ang. T. I. p.* 201. *n.* 20.

NOMADA JACOBÆÆ: nigra; abdomine fafciis fex flavis, primis tribus interruptis, antennifque flavis. *Panz. Fn. Inf. Germ. Init. n.* 72. *tab.* 20.

Found in fpring on the flowers of the goofeberry.

PLATE

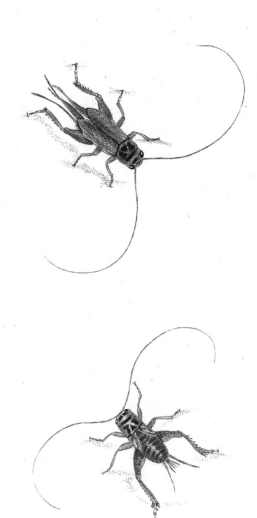

409

PLATE CCCCIX.

GRYLLUS DOMESTICUS.

HOUSE CRICKET.

HEMIPTERA.

GENERIC CHARACTER.

Head inflated, armed with jaws: feelers filiform: antennæ ufually fetaceous, or filiform: wings four, deflected, convolute, the lower ones plaited: pofterior legs formed for leaping: claws double.

** *Section* Acheta. Antennæ fetaceous: feelers unequal; thorax rounded: tail with two briftles.

SPECIFIC CHARACTER

AND

SYNONYMS.

Wings tailed, and longer than the wing-cafes: body glaucous.

GRYLLUS DOMESTICUS: alis caudatis elytro longioribus, pedibus fimplicibus, corpore glauco. *Linn. Fn. Suec.* 868.—*Scop. Ent. Carn.* 318.
Acheta Domeftica: *Fabr. Ent. Syft. T.* 2. *p.* 29. *n.* 3.
GRYLLUS DOMESTICUS: Mouffet. Inf. p. 135.
Grylli Mouffeti. *Jonft. Inf.* 12.
Le Grillon, *Geoffr. T. I. p.* 389. *n.* 2.

G 2   Fe.v

Few infects are more familiar than the common Cricket. It frequents houfes, and is fuperftitioufly efteemed by many a welcome inmate.

This little animal is not only fond of warmth, but, as though an almoft intenfe and fuffocating heat were abfolutely neceffary to its very being, it is conftantly found moft abundantly in bakehoufes, kitchen chimnies, and other places where the greateft heats prevail. Befides the inacceffibility of its lurking places in general, nothing has more fully contributed to the prefervation of thofe infects than the filly veneration which the vulgar entertain for it; interpreting its prefence as an omen of good fortune, and conceiving it would be un-propitious to harm or deftroy it.

The Cricket is indeed an animal of inoffenfive manners; it is trou-blefome only from the inceffancy of its chirping, which continues without intermiffion night and day. Some think its note louder before rain than at any other time; a circumftance afferted both by Linnæus and Fabricius. Geoffroy fays, this noife is occafioned by the friction of its thorax againft the head and wing-cafes. According to Poda, the Cricket deferts houfes infefted with the cock roach, and is deftroyed by pills of arfenic and the frefh root of the daucus mixed with flour, or the root of the nymphæa boiled in milk.

PLATE

410

2

1

1

PLATE CCCCX.

FIG. I.

APIS PICIPES.

HYMENOPTERA.

GENERIC CHARACTER.

Mouth horny: jaw and lip membranaceous at the tip: tongue inflected: feelers four, unequal, filiform: antennæ fhort, and filiform; thofe of the female fomewhat clavated: fting of the females and neuters pungent, and concealed within the abdomen.

SPECIFIC CHARACTER

AND

SYNONYMS.

APIS PICIPES. Black, covered with pale down: thorax tinged with fulvous: abdomen fufcous: legs rufous, pitchy.

MELITTA PICIPES: nigra, pallido-villofa; thorace fulvefcenti; abdomine fufco; pedibus rufo-piceis. *Kirby. Ap. Angl. T. n. p.* 127. *n.* 66.

━━━━━━

Defcribed by Mr. Kirby as a new fpecies of his genus Melitta from the individual fpecimen (Apis, n. 65, of Mr. Drury's cabinet),

figured

figured in the annexed plate. It appears, the fpecies has not been obferved in any other collection.

This infect is of a moderate fize, as the line defcribing its length at fig. 1, in the lower part of the plate is intended to fhew. The prevailing colour is black, but affumes a greyifh afpect from the pale downy hairs with which it is partially covered: the thorax is more villous than the body, and this villofity partakes in a flight degree of a fulvous tint; a few hairs of the fame colour is alfo obferveable about the head; the antennæ are blackifh; wings hyaline with the nerves pale teftaceous.

FIG. II.

APIS DISJUNCTA.

SPECIFIC CHARACTER,

AND

SYNONYMS.

Black: pofterior part of the thorax, and anterior part of the abdomen yellow downy: wings fufcous.

APIS DISJUNCTA: nigra thorace poftice abdominifque antice tomentofo flavis, alis fufcis. *Fabr. Ent. Syft. T. 2.* 328. *n.* 61.
ANTHROPHORA DISJUNCTA. *Fabr. Syft. Piez.* 374. *n.* 10.

Among the Britifh Apes in the cabinet of Mr. Drury (No. 38), we poffefs a fpecimen of this remarkable infect, and which we think

too

PLATE CCCCX.   47

too interesting to be omitted. We cannot pretend to determine on what authority it was introduced into that collection, neither are we inclined to pledge our opinion in favour of its being a genuine British Insect; we consider only that it might have been found alive in England, and under this idea may be noticed with propriety in the present work. Mr. Kirby did not consider it as an English Insect, or he would have introduced it into his Apum Angliæ. This insect, like Apis Iricolor, inserted as British in Mr. Kirby's work on the authority of a specimen in Dr. Latham's collection, is known as a native of the West Indies, and may possibly, as well as that insect, have been brought into England with some West Indian cargoes, and been afterwards discovered by accident at large in the country. Many well authenticated instances of this kind have occurred within our own knowledge. Aware of this, we cannot consistently admit an extra European insect as an aborigine, but as an occasional wanderer from the tropic regions found alive in this country; and which, from its habits of life, might even become the origin of a future British species.

We should rather suspect from the appearance of the insect, that it has been introduced in some piece of timber imported from the American islands, for it is of the same natural family as the Apis Centuncularis, or Carpenter Bee, which is well known to undergo its various transformations in centunculi, or small cells formed of leaves, and deposited in large hollow cavities bored through the timber by the parent female: this is not in the least unlikely, as this tribe of insects remain for a considerable period of time in the egg, larva, and pupa state, that the insect might have been deposited in the egg state in the West Indies, and was not liberated from its confinement till the timber, in which it was concealed, arrived in England. —The insect is shewn of its natural size in the annexed plate.

This species seems to be the insect described by Fabricius, as above referred to; and has not, we believe, been figured by any author.

PLATE

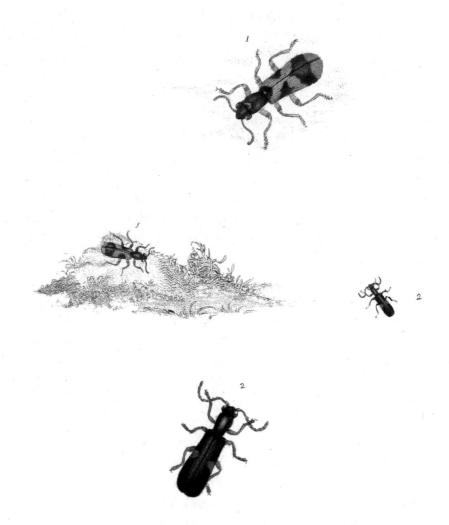

411

PLATE CCCCXL.

FIG. I. I.

CLERUS MOLLIS.

COLEOPTERA.

GENERIC CHARACTER.

Antennæ moniliform, with the three extreme joints largeft: head bent down: thorax convex and attenuated behind: wing-cafes flexile, body elongated and fomewhat oblong.

SPECIFIC CHARACTER

AND

SYNONYMS.

Grey and pubefcent, with three paler bands on the wing-cafes.

CLERUS MOLLIS: grifeus pubefcens, elytris fafciis tribus pallidis. *Marfh. Ent. Brit. T. I. p. 322. n. 2.*
ATTELABUS MOLLIS: *Linn. Syft. Nat.* 621. *n.* 11.—*Fn. Suec.* 642.
NOTOXUS MOLLIS: *Fabr. Syft. Ent.* 158. 1.—*Spec. Inf. I.* 203. *I.*—*Mant. I.* 127. *3.*—*Ent. Syft. I.* p. 211. 5.
NOTOXUS MOLLIS: *Gmel. Linn. Syft. Nat.* 1813. 3.
DERMESTES MOLLIS: *Schrank.* 37.
Curculio. *Udd. Diff.* 28. *t. I. f.* 9.
Le Chiron porte-croix. *Geoffr. I.* 305. 3.

VOL. XII.   H   A rare

A rare infe&t in England. Inhabits woods. The fmalleft figure denotes the natural fize.

FIG. II. II.

TILLUS BIMACULATUS.

GENERIC CHARACTER.

Antennæ filiform and ferrated : head fomewhat declining : thorax convex, oblong and attenuated both before and behind : body linear and elongated.

SPECIFIC CHARACTER.

TILLUS BIMACULATUS : blue-black : with a pale ferruginous fpot on the wing-cafes.

Fabricius defcribes three fpecies of the genus Tillus in his *Entomologia Syftematica*, elongatus, ambulans, & ferraticornis ; all which are found in Europe, and the two firft in this country. To thefe Mr. Marfham adds two other Britifh fpecies, æneus and virens ; including at the fame time, as a fifth fpecies, the Clerus unifafciatus of Fabricius. Thus it appears, we poffefs five defcribed fpecies of this genus in Britain, neither of which agreeing with our prefent infe&t, we prefume to offer it as a fixth and new fpecies.

Befides the five fpecies of Tillus above mentioned, the late Fabrician work, *Supplem. Ent. Syft.* contains two more fpecies, damicornis, a native of America, and Weberi, a German infe&t. Neither of thofe, however, correfpond with our infe&ts ; and we have
therefore

PLATE CCCCXI.   51

therefore no hefitation in admitting it as a non-defcript, as well as new Britifh fpecies.

We have once feen this infect alive in the woods of Kent, but accidentally loft it : the only fpecimen we now poffefs is in the cabinet of the late Mr. Drury : and we do not recollect to have ever feen it in any other.

H 2   PLATE

PLATE CCCCXII.

PHALÆNA FLEXUOSA.

YORKSHIRE Y MOTH.

LEPIDOPTERA.

GENERIC CHARACTER.

Antennæ gradually tapering from the bafe to the tip : wings in general deflected when at reft. Fly by night.

* *Noctua.*

SPECIFIC CHARACTER.

NOCTUA FLEXUOSA. Anterior wings reddifh-grey varied with fub-ferruginous : in the middle a flexuous white ftreak inclofing a pale lobiform fpot.

━━━━━━━

This fpecies of Noctua is defcribed upon the authority of an unique Britifh fpecimen in the cabinet of Mr. Drury, that has not been hitherto defcribed by any writer. It was taken in Yorkfhire, and has been denominated among Englifh collectors to whom the circumftance was known, the Yorkfhire Y Moth. We name it Flexuofa from the whitifh flexuous line, which originates at the inner edge of the firft wings near the bafe, and paffing from thence

in

in a circuitous direction to the middle of the wing, encircles a fmall lobe-fhaped fpot of a pale clay colour; in other refpects this fpecies is not unlike the common Y Moth *Phalæna Interrogationis,* but in this very ftriking character it is totally diftinct.

Our Noctua Flexuofa is very different from either of the anlagous fpecies, except that above mentioned; neither does it accord with any others defcribed in foreign entomological works with which we are acquainted. The figures which reprefent it both in an incumbent pofture, and with the wings expanded, are fufficiently correct to render any further defcription of this interefting infect unneceffary.

PLATE

413

PLATE CCCCXIII.

BANCHUS PICTUS.

HYMENOPTERA.

GENERIC CHARACTER.

Feelers four, elongated, with the joints cylindrical : **lip at the** bafe cylindrical and horny, tip membranaceous, rounded, **and** entire : antennæ fetaceous.

SPECIFIC CHARACTER

AND

SYNONYMS.

Black varied with yellow; fcutel fomewhat fpinous.

BANCHUS PICTUS: niger flavo varius fcutello fubfpinofo. *Fabr.* *Supp. Ent. p.* 234. *n.* 7.

———————

Defcribed by Fabricius as a native of Germany from the cabinet of Smidt.—Not before noticed as a Britifh fpecies.

The fmalleft figure denotes the true fize of this curious infeft.

PLATE

PLATE CCCCXIV.

FIG. I. I.

CURCULIO VAU.

COLEOPTERA.

GENERIC CHARACTER.

Antennæ clavated, and feated on the fnout, which is horny and prominent: pofterior part of the head thick.

SPECIFIC CHARACTER

AND

SYNONYMS.

Wing cafes with a fpot, and common V-mark of white.

CURCULIO VAU: elytris macula et figura V communi albis. *Marſh. Ent. Brit. T.* 1. *p.* 299. *n.* 177.—*Schrank,* 227.—*Vill. I.* 214. 174.

—————

The body of this infect is rather downy, and of a cinereous colour; the figure fomewhat oblong, and the fnout thick. The common V-like mark, formed by the union of two oblique whitifh lines on the pofterior part of the future, and the fingle anterior white fpot in the middle of each of the wing-cafes, are very characteriftic of this fpecies.—The fmaller figure denotes the natural fize.

VOL. XII.   I   FIG.

PLATE CCCCXIV.

FIG. II. II.

CURCULIO EXARATUS.

SPECIFIC CHARACTER.

Fufcous: wing-cafes cinereous, and rather deeply marked with dotted ftriæ.

CURCULIO EXARATUS: fufcus, elytris cinereis profundiufcule ftriatis: ftriis punctatis. *Marfh. Ent. Brit. T.* 1. *p.* 303. *n.* 188.

———————

The fnout is rather broad and thick: wing-cafes pale fufcous, or cinereous, and marked with moderately deep dotted lines: the whole of the under furface covered with very fhort down.

———————————————

FIG. III. III.

CURCULIO FASCIATUS.

SPECIFIC CHARACTER

AND

SYNONYMS.

Ferruginous-fufcous: wing-cafes fafciated with white: legsrufous.

CURCULIO

PLATE CCCCXIV. 59

CURCULIO FASCIATUS: fufco-ferrugineus, elytris albo-fafciatis, pedibus rufis. *Marfh. Ent. Syft. T. I. p.*286. 144.

Curculio Incurvus? *Panz. Faun. Germ.* 36. *t.* 7.

―――――

Found on the leaves of the Cratægi Oxyacanthe.

The antennæ, fnout, and legs rufous : thorax ferrugineous-brown, with a thin white dorfal line : fcutel white : wing-cafes ferruginous brown and ftriated; the ftriæ impreffed with dots.—When in fine prefervation, this is an elegant fpecies, being variegated, though in a flight degree only, with white, befides the broad whitifh commonb and towards the pofterior part of the wing-cafes.

I 2   PLATE

PLATE CCCCXV.

LIBELLULA ÆNEA.

BRASSY-GREEN DRAGON-FLY.

NEUROPTERA.

GENERIC CHARACTER.

Mouth armed with more than two jaws: lip trifid: antennæ
fhorter than the thorax, very thin, and filiform: wings expanded:
tail of the male furnifhed with a furcated procefs.

SPECIFIC CHARACTER

AND

SYNONYMS.

Wings tranfparent: thorax braffy-green.

LIBELLULA ÆNEA: alis hyalinis, thorace viridi æneo. *Linn.*
        Fn. Suec. 1466.—*Gmel. Linn. Syft. Nat.* 2622.
        n. 8.
        Fabr. Spec. Inf. I. p. 524. *n.* 27.
Libellula viridi aurata capite rotundato, pedibus nigris, abdominis
        medio inflato. *Degeer. Inf.* 2, 2. 52. *tab.* 19.
        fig. 8.
Libellula viridi nitens, alis pallidis, pedibus nigris. *L'Amianthe.*
        Geoffr. Inf. 2. 226. 10.
        Roef. Inf. 2. *aqu.* 2. *t.* 5. *f.* 2.
        Schaeff. Ic. t. 113. *f.* 4.
        Raj. Inf. p. 49. *n.* 5.

                                            Libellula

Libellula Ænea has been recently obferved in fwampy grounds in the neighbourhood of Hampftead. It is recorded as a Britifh Infect by Ray; but fince his time appears to have become very fcarce till lately, when a fmall number of them were obferved in the above-mentioned fituation. We poffefs two varieties taken in this place, in the fummer of 1805, which differ in fome flight particulars only. Both infects have the eyes of a brown colour, and the thorax of a brilliant green with a braffy luftre: the principal difference confifts in the colour of the wings, which in one fpecimen are hyaline, while on the contrary the wings of the other are tinged with teftaceous yellow. This yellow-winged variety is further diftinguifhed by having the body gloffed with fine golden purple; in the other, the body is of a braffy green colour, with only a flight inclination to reddifh brown*. Both infects are figured in the annexed plate in their natural fize.

This fpecies is not peculiar to England. Linnæus defcribes it as a native of Sweden; from Geoffroy we learn, that it is a native of France; and from Roefel, as being found in Germany.

----

* An indifferent figure of a fpecies of Libellula, much refembling this, occurs in the work of Schaeffer, *Icon. Ratif. pl.* 167, *fig.* 4, and which is probably intended for our yellow-winged variety.

PLATE

PLATE CCCCXVI.

VESPA CRIBRARIA.

HYMENOPTERA.

GENERIC CHARACTER.

Mouth horny, with a compreſſed jaw: feelers four, unequal and filiform: antennæ filifiorm, the firſt joint longeſt and cylindrical: eyes lunar: body glabrous: ſting pungent, and concealed within the abdomen: upper wings folded in both ſexes.

SPECIFIC CHARACTER

AND

SYNONYMS.

Black: abdomen banded with yellow, the middle ones interrupted: anterior ſhanks with concave ſhields.

VESPA CRIBRARIA: *Linn. Syſt. Nat.* 12. 2. *p.* 945. *n.* 23.—*Fn. Suec.* 2. *n.* 1675.

CRABRO CRIBRARIA: nigra, abdomine faſciis: intermediis inteṛ-ruptis, tibiis anticis clypeis concavis. *Fabr. Sp. Inſ. I. p.* 470. *n.* 8.—*Mant. Inſ. I. p.* 296. *n.* 13.

Apis tibiis anticis lamella cribriformi. *Uddm. Diſſ.* 94.

Ray Inſ. p. 255. *n.* 15.

Rolander Aɛt Stockh. 1751. *p.* 56. *t.* 3. *f.* 1.

Crabro Cribrarius: *Panzer Fn. Inſ. Germ.*

Found

Found in England, in Sweden, and in Germany. The male has the fhanks of the anterior legs fhielded, while in the female thofe parts are fimple: this characteriftic of the two fexes is not peculiar to our infect, we obferve the fame in the Fabrician *Crabro Clypeatus* (Vefpa Clypeata, *Gmel.*), in Crabro Scutatus (Vefpa Scutata, *Gmel.*), and feveral other fpecies of the Linnæan Vefpæ.

PLATE

margin of the wing-cafes is another : near the future, and before the middle of the wing-cafes, is a fourth fpot; and a fifth towards the end, placed tranfverfely.—Varieties occur in which the thorax is reddifh at the fides.

FIG. II.

SCARABÆUS SPHACELATUS.

SPECIFIC CHARACTER.

Black : head tuberculated : margin of the thorax pale: wing-cafes grifeous with dotted ftriæ, and a fingle fufcous daub or irregular fpot.

SCARABÆUS SPHACELATUS : niger, capite tuberculato, margine thoracis pallido, elytris grifeis punctato-ftriatis : litura unica fufca. *Marfh. Ent. Brit. T. I. p. 15, n. 20.*
Panz. Faun. Germ. 58. t. 5.

Very abundant in dung. Size the fame as in the preceding fpecies. The antennæ of this infect is black: head obfolete, tuberculated, and black: thorax black, gloffy, and very minutely punctured, with the lateral margin livid: fcutel fufcous: body black beneath : legs pale.

FIG.

PLATE CCCCXVII. 67

FIG. III. III.

SCARABÆUS FOSSOR.

SPECIFIC CHARACTER,

AND

SYNONYMS.

Black: thorax fomewhat retufe: head with three tubercles and fomewhat cornuted in the middle.

SCARABÆUS FOSSOR: niger, thorace fubretufo, capite tuberculis tribus: medio fubcornuto. *Linn. Syft. Nat.* 548. 31.—*Fn. Suec.* 384.—*Fabr. Sp. Inf. I. p.* 15. *n.* 59.—*Mant. Inf. I. p.* 8. *n.* 62.—*Marfh. Ent. Brit. T. I. p.* 16. *n.* 24.
La Tete Armée: *Geoffr. Inf. I. p.* 82. *n.* 20.
Schaeff. Icon. t. 144. *f.* 78.

———————

The length of this infeƈt rather exceeds three-eights of an inch: the colour is entirely black, glabrous, fmooth, and fhining; its thorax is very convex: wing-cafes oblong and ftriated: antennæ lamellated; and wings fufcous. Found in dung.

K 2   FIG.

FIG. IV.

SCARABÆUS RUFIPES.

SPECIFIC CHARACTER

AND

SYNONYMS.

Pitchy: antennæ pale: wing-cafes fmooth.

SCARABÆUS RUFIPES : piceus, antennis pallidis, elytris lævibus.
      Linn. Syft. Nat. 559. 86.—*Fn. Suec.* 403.—
      Gmel. 1552. 86.—*Marfh. Ent. Brit. T. I.*
      p. 25. *n.* 42.
SCARABÆUS CAPITATUS : *De Geer,* 4. *p.* 263. 7. *t.* 10. *f.* 6.
SCARABÆUS OBLONGUS : *Scop.* 19.

———

About the fize, and has the fame habits as the preceding, the
figure in the annexed plate being magnified. It is entirely of a black
colour and gloffy: fhield of the head obtufe: laft joints of the legs
pale rufous.

PLATE

PLATE CCCCXVIII.

SCARABÆUS GREENII.

GREEN'S SCARABÆUS.

COLEOPTERA.

GENERIC CHARACTER.

Antennæ clavated: the club fiffile : fhanks of the anterior legs generally dentated.

SPECIFIC CHARACTER

AND

SYNONYMS.

Blackifh : thorax and margin of the wing-cafes fprinkled with white dots: on the firft four fegments of the abdomen beneath a fingle white dot in the middle, and one each fide.

CETONIA VARIEGATA: atra thorace margine elytrifque maculis albis fparfis. *Fabr. Ent. Syft. T. I. p.* 2. 151. *n.* 88 ?

Scarabæus tribus antennarum lamellis tricefimus quintus. *Schaef. Icon. pl.* 198. *fig.* 8 ?

SCARABÆUS ALBELLUS: *Pallas. Icon. I.* 17. *tab. A.* 18 ?

━━━━━━

Among the Englifh Scarabæi in the cabinet of the late Mr. Green, we poffefs a fingle fpecimen of this very interefting fpecies. It is

of

of the genus Scarabæus in the Linnæan arrangement; Cetonia of Fabricius. We are totally unacquainted with its hiftory, and, believing it to be unnoticed by any former writer, have named it after its former poffeffor, S. Greenii.

It fhould be particularly obferved, that this infect approaches very clofely to the Cetonia Variegata of Olivier and Fabricius, which is a native of Tranquebar. We fhall not prefume to fay, that it may not be a variety of that variable fpecies; but we certainly think it is not, if the general defcription the works of Olivier afford us be correct. The upper furface of our infect nearly correfponds with his figure and defcription; the principal difference confifts in the number and difpofition of the white fpots on the lower furface of the abdomen, and thofe are ftrikingly diffimilar. Olivier defcribes his fpecies as having two rows or lines of white dots on each fide, while in our infect there is only one on each fide; and the fingle row of white dots down the center in our infect does not agree exactly with Cetonia Variegata. We have previoufly ftated, that the upper furface of the two infects are fimilar, but we fhould further add, that they are not perfectly fo, as Olivier's infect has the pofterior border of the thorax white; and in our infect, that part differs in no refpect from the reft of the thorax in point of colour, except the fpots, which are white and impreffed.—Our fynonyms refer to the figure of an European fpecies of Scarabæus in the works of Schaeffer, that feems to bear a general refemblance to our infect; yet we muft confefs, we can only entertain a very remote idea of its being intended for our infect; the indifference of the figure precludes the poffibility of determining this circumftance with accuracy: it is about the fame fize, of a blackifh colour and fpotted with white; but there is alfo an appearance in the figure of the infect being hairy, and if that be correct, it cannot be the fame as our infect.

The fmalleft figure denotes the natural fize of this infect. The two other figures reprefent the upper and lower furfaces of the infect magnified.

PLATE

1

PLATE CCCCXIX.

MUSCA GROSSIFICATIONIS.

DIPTERA.

GENERIC CHARACTER.

Mouth with a soft, exserted fleshy proboscis, and two unequal lips: sucker beset with small bristles: feelers short, and two in number, or sometimes none: antennæ usually short.

SPECIFIC CHARACTER

AND

SYNONYMS.

Deep black: wings black, tipped with white.

Musca Grossificationis: atra, alis nigris apice albis. *Linn. Fn. Suec.* 1865.

Musca antennis setariis alis nigris apice albis. *Linn. Syst. Nat. Edit.* 10. *p.* 599. *n.* 84.

Musca nigra alis fuscis, apicibus albis. *Act. Upsf.* 1736. *p.* 33. *n.* 50. *Fabr. Sp. Insf.* 2. *p.* 451. *n.* 83. *Gmel. Linn. Syst. Nat.* 2855. 109.

La mouche à aîles noires & tache blanche à l'extrémeté. *Geoff. Insf.* 2. *p.* 493. *n.* 1.

This

This diminutive fpecies of Mufca is mentioned by Linnæus as one of the rareft of all the European infeĉts. It is certainly uncommon, and efpecially in England. Geoffroy found it on flowers in the royal garden at Paris. The fmalleft figure denotes the natural fize.

PLATE

420

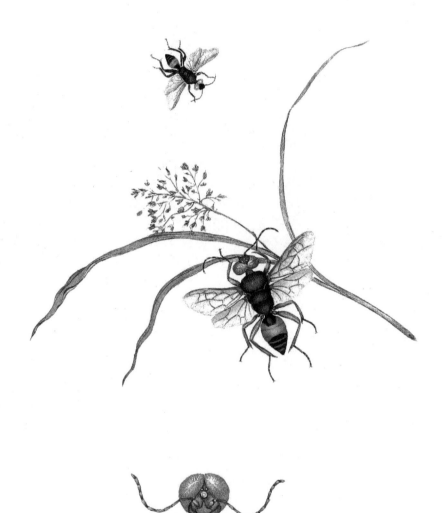

PLATE CCCCXX.

LARRA POMPILIFORMIS.

HYMENOPTERA.

GENERIC CHARACTER.

Tongue porrected, fimple : jaws fhort, horny, vaulted : lip exferted, membranaceous at the tip and marginated : antennæ filiform.

SPECIFIC CHARACTER

AND

SYNONYMS.

Black: abdomen black, with the bafe ferruginous.

Larra Pompiliformis : nigra, abdomine nigro bafi ferrugineo.— Die Grabwefpenartige Drehwefpe. *Panz.* 89. *n.* 13.

———————

A curious little infect of the Fabrician genus Larra. The only fpecimen we have yet feen of this infect occurs in the cabinet of Mr. Drury. The fmalleft figure explains the natural fize.

VOL. XII.   L   PLATE

PLATE CCCCXXI.

FIG. I. I.

APIS SIGNATA.

HYMENOPTERA.

GENERIC CHARACTER.

Mouth horny: jaw and lip membranaceous at the tip: tongue inflected: feelers four, unequal, filiform: antennæ fhort, and filiform; thofe of the female fomewhat clavated: fting of the females and neuters pungent, and concealed within the abdomen.

SPECIFIC CHARACTER

AND

SYNONYMS.

Apis Signata: black; front of the head, and bafe of the thorax yellow.

Sphex Signata: atra, nitida, immaculata; alis albis; fronte fub antennis maculis duabus flavis. *Panz. Fn. Inf. Germ. Init. n. 53. t. 2.*

Vespa: nigra, fronte, thoracifque bafi flavis.—La guêpe noire, à levre fuperieure & bafe du corcelet jaunes. *Geoffr. Inf. n. p. 379. n. 14.*

Vespa Pratensis: *Fourcroy. Ent. Par. n. 14.*

Melitta Signata: atra; fronte maculata; abdomine fegmento primo margine utrinque albo.—Var. γ antennis fubtus piceis, collare tuberculifque luteis. *Kirby, Ap. Ang. T. 2. p. 41.*

L 2

This

This little infe&ct; is found on different fpecies of Refeda, or mignonet, and in common with various other infe&cts; of the fame natural family, that are ufually found on thofe plants, emit a ftrong odoriferous fcent. —There are two or more varieties of this infe&ct;: that figured by Panzer under the name of Sphex Signata, has two yellow fpots on the front of the head, inftead of the whole fpace beneath the antennæ being yellow, as in our fpecimen. The fmalleft figure denotes the natural fize.

FIG. II.

APIS LÆVIGATA β.

SPECIFIC CHARACTER

AND

SYNONYMS.

Black, and fomewhat pubefcent with rufous hairs : thorax ferruginous: abdomen highly polifhed, with the three middle fegments, pale each fide at the bafe.

MELLITTA LEVIGATA : atra rufo-fubpubefcens ; thorace ferrugineo ;
         abdomine nitidiffimo, fegmentis intermediis bafi
         pallefcentibus.   *Kirby, Ap. Ang.* 2. 75. 32.
         var. β.

This is an elegant little fpecies, and very rare. The figure reprefents it in the natural fize.

FIG.

FIG. III.

APIS OCHROSTOMA.

SPECIFIC CHARACTER

AND

SYNONYMS.

Deep black: fcutel fanguineous : abdomen rufous, variegated with fpots and interrupted bands of yellow.

Apis Ochrostoma : atra : fcutello fanguineo; abdomine rufo, maculis fafciifque interruptis, flavis, variegato. *Kirby Ap. Angl. T. 2. p. 209. n. 26.*

Defcribed by Mr. Kirby as a new fpecies of Apis, from a fpecimen in the cabinet of Mr. Drury.

PLATE

422

PLATE CCCCXXII.

PTINUS FUR.

COLEOPTERA.

GENERIC CHARACTER.

Antennæ filiform, the exterior joint largeft: thorax fubrotund, without margin, and receiving the head.

SPECIFIC CHARACTER

AND

SYNONYMS.

Ferruginous brown: thorax four-toothed: wing-cafes with two white bands.

PTINUS FUR: fufco-ferrugineus, thorace quadridentato, elytris fafciis duabus albis. *Marfh. Ent. Brit. T. I. p.* 89. *n.* 27.

PTINUS FUR: teftaceus fubapterus, thorace quadridentato, elytris fafciis duabus albis. *Fabr. Spec. Inf. p.* 73. *n.* 4.—*Mant. Inf. I. p.* 40. *n.* 4.—*Ent. Syft. I.* 2. 39. 4.—*Gmel. Linn. Syft. p.* 1607. *n.* 5.

CERAMBYX FUR.—*Linn. Fn. Suec.* 651.
Preys. Boh. Inf. 56. 57.
Stroem. Aĉt. Nidrof. 111. 393. 12.

BRUCHUS TESTACEUS: La Bruche à bandes. *Geoffr. Inf. Parif. I. p.* 164. *n.* 4. *t.* 2. *f.* 6.

PTINUS

Ptinus Rapax : *Degeer. Inf. 4. p. 231. n. 5. t. 9. f. 5. 6. 7.*
Buprestis Fur : *Scop.* 210.

━━━━━━━━━━

This deſtructive little infect is produced from a ſoft and hairy fix-footed larva of a ferruginous colour, which preys on furniture, books, and other ſimilar articles. It reſides principally in wood, occupying ſmall tubular cavities, which it perforates in a variety of directions, reducing, as it proceeds in its devious courſe, the hardeſt timber within the dimenſions of its receptacle to a light duſt or powder. Ptinus Fur is alſo very detrimental to preſerved articles of natural hiſtory. The nymph or pupa is contained in a glutinous follicle.

PLATE

PLATE CCCCXXIII.

LIBELLULA FORCIPATA.

FORCIPATED DRAGON-FLY.

NEUROPTERA.

GENERIC CHARACTER.

Mouth armed with more than two jaws: lip trifid: antennæ fhorter than the thorax, very thin and filiform: wings expanded: tail of the male furnifhed with a furcated procefs.

SPECIFIC CHARACTER

AND

SYNONYMS.

Thorax greenifh yellow, with black lines: abdomen blackifh with yellow charaćters.

Libellula Forcipata : thorace luteo-virefcente, lineis nigris;
abdomine nigricante charaćteribus flavis. *Linn.*
Fn. Suec. n. 771.
Gmel. Linn. Syft. Nat. 2625. *n.* 11.
Æshna Forcipata : thorace nigro : charaćteribus varius flavef-
centibus, cauda unguiculata. *Fabr. Ent. Syft.*
T. 2. *p.* 383. *n.* 1.
Libellula nigra capite rotundato, thorace fegmentifque aliquot abdo-
minis viridi maculatis. *Degeer. Inf.* 2. 2. 50.
Libella major, corpore compreffo flavefcente. *Petiv. Muf.* 84.
n. 819.

VOL. XII.   **M**   Libella

Libella maxima lutea, cùm 4 vel 5 ſpinis in extremitate caudæ.
Merret Pin. 197. *n.* 4.
La Caroline.   *Geoffr. Inſ. t.* 2. *p.* 228. *Sp.* 15.

———

This is an intereſting, rare, and elegant ſpecies of Libellula. In
the cabinet of the late Mr. Drury, we have a ſingle ſpecimen of this
ſcarce inſect; another has been recently taken near Highgate, and is
likewiſe in our poſſeſſion. Like the reſt of its tribe it is found in
ſwamps and other watery places. The larva is unknown to us, its
pupa is repreſented with the perfect inſect in the annexed plate.

The head of this ſpecies of Libellula is of a fine yellow, faciated
with black: the eyes prominent, brown, and gloſſy: the thorax
greeniſh yellow, lineated with black; the abdomen black with a longitudinal interrupted dorſal line of whitiſh yellow, and the middle
ſegments marked on both ſides with a ſhort tranſverſe yellowiſh band,
and a ſemi-lunar mark of the ſame colour below it: the wings tranſ-
parent, with a dark anterior coſtal mark, as in moſt other of the
Libellula tribe

PLATE

Less than the common house-fly, *Musca Domestica.* It inhabits various parts of Europe, and is not common in England.

The thorax is of a brownish colour lineated with dusky : the body nearly round, and marked down the middle of the back with a series of blackish spots or dots ; the wings are whitish, faintly tinged with testaceous towards the base, and the legs blackish.

FIG. II.

MUSCA ARCUATA.

SPECIFIC CHARACTER

AND

SYNONYMS.

Elongated, black, spots on the sides of the thorax, and four arcuated bands on the abdomen yellow.

Musca Arcuata : nigra, antennis elongatis, thorace maculis la-
teribus, abdomine cingulis quatuor arcuatis flavis.
Linn. Fn. Suec. 1806.
Syrphus Arcuatus : *Fabr. Ent. Syst. T.* 4. *p.* 293. *n.* 55.
Panz. Fn. Germ. 2. *tab.* 10.

This species is found on flowers. It is a general inhabitant of Europe; in England this insect is scarce.

Fabricius describes a very distinct species from this under the name of Musca Arcuata; the insect we have figured is the Musca Arcuata only of Linnæus, Syrphus Arcuatus of Fabricius.

PLATE

PLATE CCCCXXV.

LIBELLULA QUADRIFASCIATA.

NEUROPTERA.

GENERIC CHARACTER.

Mouth armed with more than two jaws; lip trifid: antennæ shorter than the thorax, very thin and filiform: wings expanded: tail of the male furnished with a furcated procefs.

SPECIFIC CHARACTER.

LIBELLULA QUADRIFASCIATA: wings white, tinged anteriorly with yellowish; tips of all the wings, and bafe of the pofterior pair with a fufcous band.

———

This evidently new fpecies of Libellula occurs in the cabinet of Mr. Drury: it correfponds in fome refpects with Libellula Quadri-maculata (See Plate 407), the fize and general afpect of both infects is nearly the fame, but the prefent fpecies differs among other particulars in having a fufcous band at the tip of all the wings. Libellula Rubicunda alfo bears fome refemblance to this infect, except that the tips of the wings are perfectly immaculate, as in Libellula 4-maculata.—Our new fpecies Libellula Quadrifafciata is a very rare infect, and has not been noticed by any author.

PLATE

PLATE CCCCXXVI.

PAPILIO BLANDINA.

SCOTCH ARGUS BUTTERFLY.

LEPIDOPTERA.

GENERIC CHARACTER.

Antennæ terminated in a club : wings erect when at rest : fly by day.

SPECIFIC CHARACTER

AND

SYNONYMS.

Wings indented, fuscous, with an ocellar rufous band : posterior pair beneath fuscous, with a cinereous band.

Papilio Blandina: alis dentatis fuscis: fascia rufa ocellata posticis subtus fuscis: fascia cinerea. *Fabr. Ent. Syst. T. 3. p. I.* 236. *n.* 736.

This very rare species of Papilio has been recently discovered to be a native of the British isles. About three or four specimens of it were taken in the isle of Arran by Major Walker, to whose politeness we are indebted for the individual example at this time in our Museum. Another is preserved in the collection of our friend A. M'Leay, Esq. and those, we have reason to apprehend, are the only specimens at present in any of the London Cabinets.

Though but lately introduced to our attention as a native of Great Britain, this interesting insect is by no means unknown to the continental

tinental naturalifts as an inhabitant of Germany. It is the true
Papilio (*Sat.*) Blandina of the Fabrician fyftem *. This author
likewife defcribes another Papilio, nearly allied to the above, under
the fpecific name of Ligea. This latter is, however, fufficiently
diftinguifhed by having four ocellate black fpots in the rufous band
on the upper wings inftead of three, as in P. Blandina. Fabricius,
in his general defcription, fpeaks of the near affinity his P. Blandina
bears to P. Ligea, but obferves that P. Ligea has a white fpot at the
end of the band on the underfide of the pofterior wings, which the
other has not. " Affinis P. Ligea. differt tamen alis pofticis vix
ocellatis, fufcis fafcia cinerea abfque maculis albis." *Fabr.*—Papilio
Ligea was difcovered by Major Walker in the ifle of Arran at the
fame time as P. Blandina, and will fhortly appear in the prefent
work †.

---

* Fabricius defcribes two of the Papiliones under the fpecific name of Blandina, but
which cannot eafily be confounded, as one of them are of the *Pap. Nymphales* tribe, and
the other belongs in his arrangement to the *Satyri.*—P. N. Blandina is an Eaft Indian
fpecies, and is fully noticed in our illuftration of Exotic Entomology.

† Figures of both the above-mentioned infects have appeared in a late publication,
the " Britifh Mifcellany," one in Plate 2, the other in Plate 7. Unfortunately, however, the Editor has entirely mifconceived the Fabrician authorities, and reverfed the two
names affigned them by that author. The Fabrician Papilio Blandina is by that means
erroneoufly named Ligea, and, *vice verfa*, the Fabrician P. Ligea, called Blandina.—
There are, befides, a few errors in the figures with regard to the form, fituation, and
number of the ocellate fpots. Thofe relating to the P. Ligea will be hereafter noticed.
In our P. Blandina (Ligea *Brit. Mifc.*) the macular band on the underfide of the pofterior
wings appears to have only two fmall dots, while in the infect there are no lefs than fix,
the three lower of which has a white dot in the center.—We were at firft inclined to fufpect,
that the fpecimen in the cabinet of Mr. M'Leay, from which the drawing of that infect
was taken, might have been in fome meafure injured, and the fpots obliterated, or that
his infect varied from that we poffefs; we have, however, fince compared them, in order
to afcertain whether any fuch diffimilarity in reality exifted between them, and find
the two infects correfpond in every refpect.

PLATE

PLATE CCCCXXVII.

FIG. I. I.

ARANEA LIVIDA.

APTERA.

GENERIC CHARACTER.

Mouth with short horny jaws: lip rounded at the tip: feelers two, incurved, jointed, and acutely pointed, those of the male clavated and furnished with the sexual organs: antennæ none: eyes eight or rarely six: legs eight: papillæ for spinning at the tip of the abdomen or vent.

Section eyes ..: :..

SPECIFIC CHARACTER.

ARANEA LIVIDA: thorax subtriangular: abdomen ovate; above brown, obscurely dotted, and lineated with blackish: beneath testaceous.

———

A specimen of this remarkable species of Aranea occurs in the cabinet of the late Mr. Drury, with a memorandum relating the following particulars of its capture.—" This spider was taken out of the water at Hornsey wood, October 4th, 1766, being in company with Mr. Rice."

VOL. XII.   N   The

The figures, which fhew both the upper and lower furface of this gigantic fpider, fufficiently exemplifies its magnitude and general afpect, being reprefented in its natural fize. The prevailing colour of the upper furface is darker than the lower; it is a livid brown faintly variegated with reddifh. On very clofe infpection, the thorax appears to be obfcurely lineated and dotted with blackifh, radiating from the ridge of the back, as from a center towards the outer margin: the legs alfo are lineated with about four or five equidiftant blackifh lines fprinkled with a few dots, flightly hairy, and fparingly befet with fmall fetiform fpines: the abdomen rather downy.

The eyes of this fpider, eight in number, are difpofed on the anterior part of the thorax in a fingular manner: the four anterior ones form a tranfverfe curved line, behind which are two contiguous eyes of a fimilar fize, and a little farther behind two more; but the laft are placed much more remotely from each other than the former. Thofe pofterior eyes are diftinguifhed likewife by being ftationed each upon the fummit of a rather large fmooth lateral tubercle of a rufous colour. The exterior eye on each fide in the anterior line, it fhould be obferved, is feated on a fimilar fmooth rufous tubercle, but which is of a diminutive fize compared with thofe on which the pofterior eyes are fituated.

As there is no fpecies of the family to which this fpider belongs among thofe already defcribed by Linnæus, Fabricius, or any other entomological author within our knowledge, that correfponds with our prefent infect, we confider it as a new fpecies.

FIG.

FIG. II.

ARANEA MARGINATA.

SPECIFIC CHARACTER.

ARANEA MARGINATA: brown: thorax and abdomen furrounded
 with a whitifh line.
ARANEA PALUSTRIS: *Linn. Syft. Nat.* 12. 2. *p.* 1036. *n.* 41.
 —*Var.*?
ARANEA TRILINEATA: *Fabr. Ent. Syft. T.* 2. *p.* 423. *n.* 61?
Aranea pugnax: *Rofs. Fn. Etr.* 2. 135. 980?

━━━━━━━━

This infe&t agrees very nearly with the Aranea trilineata* of
Fabricius, and does not appear very remote from the Linnæan defcription of Aranea paluftris†. The principal difference confifts in
the pofition of the eyes, which conftitutes an effential chara&teriftic
mark of the feveral families into which the Aranea genus is divided;
and in this particular they are very diftin&t. Admitting therefore, that
Linnæus and Fabricius are corre&t in defcribing the fituation of the
eyes in the two infe&ts above-mentioned, we muft confider the prefent
infe&t as a fpecies diftin&t from either, notwithftanding their fimilarity
in other refpe&ts. It does not certainly agree with the defcription of

---

* *Aranea Trilineata:* fufca thoracis margine linea dorfali margineque ovato cinereis.
 Fabr. Ent. Syft. T. 2. *p.* 423. *n.* 61. *oculis* ⋮ ⋮

† Aranea Paluftris: fufca, thorace abdomineque utrinque linea nivea. *Linn. Syft.*
 Nat. 12. 2. *p.* 1036 *n.* 41.

N 2   any

any fpecies of Aranea in the fame family hitherto defcribed by thofe writers. It may therefore be a nondefcript fpecies, though we muft at the fame time confefs, we conceive it not unlikely, that the pofition of the eyes in the Aranea trilineata of Fabricius has been miftaken by that writer, and that it may hereafter prove to be the fame fpecies.

PLATE

4

3

1

2

PLATE CCCCXXVIII.

COCCINELLA 13-MACULATA,

13-SPOT LADY COW.

COLEOPTERA.

GENERIC CHARACTER,

Antennæ clavated, club folid: anterior feelers femicordated: thorax and wing-cafes margined: body hemifpherical: abdomen beneath flat.

SPECIFIC CHARACTER
AND
SYNONYMS.

Wing-cafes yellow, with thirteen black dots: body orbicular.

COCCINELLA 13-MACULATA: coleoptris flavis: punctis nigris tredecim corpore orbiculato. *Marfh. Ent. Brit. T. I.* 157. *Fabr. Syft. Ent.* 83. 24.—*Sp. Inf. I.* 99. 37.—*Mant. I.* 58. 53.—*Ent. Syft. I. a,* 279. 60.
Gmel. 1652. 90.

═══════

Three varieties of this elegant fpecies of Coccinella are figured on the fame plate, two of which differ only in colour, the third in the dots on the wing-cafes. The thorax in all the fpecimens are pale yellow,

yellow, but in the fecond individual, the wing-cafes are of a deeper orange than ufual; and in the third, the two inner dots of the three, which conftitute the macular feries acrofs the middle of the wing-cafes, are united, and form but a fingle confluent fpot. The fmalleft figure denotes the natural fize.

PLATE

PLATE CCCCXXIX.

MUSCA HEMIPTERUS.

DIPTERA.

GENERIC CHARACTER.

Mouth with a foft exferted flefhy probofcis, and two unequal lips: fucker befet with briftles: feelers fhort, and two in number, or fometimes none: antennæ ufually fhort.

* Syrphus: antennæ naked.

SPECIFIC CHARACTER

AND

SYNONYMS.

Musca Hemipterus. Downy: thorax with a ferruginous border: wings thick, cinereous: fufcous varied with yellowifh.

Syrphus Hemipterus: antennis fetariis tomentofus thoracis limbo ferrugineo, alis craffioribus cinereis fufco flavefcentique variis. *Fabr. Ent. Syft. T. 4. 284, n. 22.*
Schaeff. Icon. Tab. 71. fig. 6?
Musca Subcoleoptrata: *Gmel. Linn. Syft. Nat. 2869. n. 335.*

An infect nearly allied to the Linnæan *Conops fubcoleoptrata,* or *Mufca fubcoleoptrata* of the Gmelinian edition of the Syftema Naturæ.

Gmelin

Gmelin confiders it as the fame infect, but Fabricius defcribes it as a diftinct fpecies, diftinguifhing the *M. Subcoleoptrata* by the black thorax, and cinereous wings with two brown flexuous ftripes ; and the *M. Hemiptera* (our prefent fpecies) by the thorax being furrounded by a ferruginous border, and the cinereous wings varied with yellowifh.—Fabricius has fince conftituted a new genus of Dipterous infects, under the name of THEREVA, in which both thefe fpecies of Mufcæ are included *.

This infect is very uncommon in England.

---

* *Suppl. Ent. Syft. I. p.* 560. *n.* 2.

PLATE

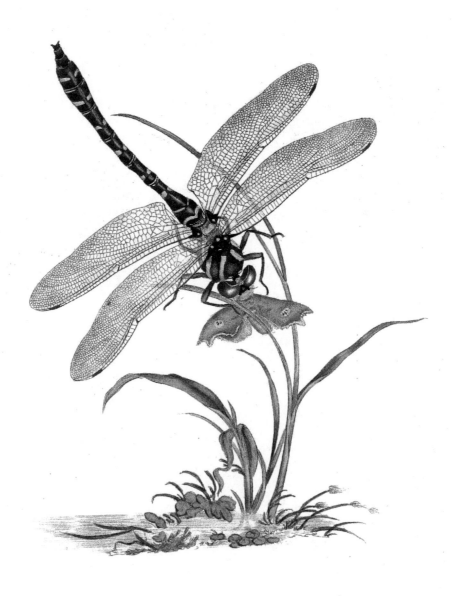

PLATE CCCCXXX.

LIBELLULA BOLTONII.

BOLTON's DRAGON-FLY.

NEUROPTERA.

GENERIC CHARACTER.

Mouth armed with more than two jaws: lip trifid: antennæ fhorter than the thorax, very thin and filiform: wings expanded: tail of the male furnifhed with a furcated procefs.

SPECIFIC CHARACTER.

LIBELLULA BOLTONII: wings hyaline: body elongated, black, with a larger interrupted yellow band acrofs the middle, and a fmaller near the tip of each feg- ment.

━━━━━━━━━

This fine and noble fpecies of Libellula appears to be unknown to any of the entomological writers we are acquainted with. The fpecimen from which our figure in the annexed plate is taken, was difcovered in Yorkfhire fome years ago by Mr. Bolton, and commu- nicated to Mr. Drury, in whofe cabinet it has remained unnoticed till the prefent time. We believe this fpecimen to be unique, or at leaft we have never feen an other. It is a beautiful, large, and in- terefting fpecies, and poffeffes characters fo extremely different from any of the known fpecies of its genus, that it cannot eafily be mif- taken. We name it Boltonii, in compliment to Mr. Bolton, the gentleman to whom we are indebted for its difcovery.

VOL. XII.   O   PLATE

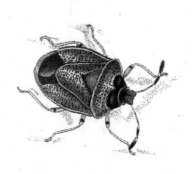

PLATE CCCCXXXI.

CIMEX MELANOCEPHALUS.

HEMIPTERA.

GENERIC CHARACTER.

Snout infle&ted: antennæ longer than the thorax: wings four, folded crofs-wife, anterior part of the upper pair coriaceous: back flat: thorax margined: legs formed for running.

SPECIFIC CHARACTER

AND

SYNONYMS.

Grey: head, and fcutel at the bafe braffy black.

Cimex Melanocephalus: grifeus capite fcutelloque bafi nigro aeneis. *Fabr. Ent. Syft. T.* 4. *p.* 125. *n.* 176.

Fabricius appears to be the only writer who has noticed this elegant little fpecies of Cimex: he defcribes it as an Englifh Infe&t, without referring to any cabinet.

This infe&t is of a fmall fize, as fhewn by the central figure in the annexed plate. The upper furface is of a greyifh colour, tinged in the fhades with green; the head, two confluent fpots at the anterior part of the thorax, and triangular fpot at the bafe of the fcutel, braffy

O 2   black,

black, inclining to purplifh. The whole furface is minutely punctured. Beneath, the prevailing colour is braffy blackifh purple, with
the margin of the abdomen whitifh, and marked with a row of
black dots: legs yellowifh, with a black dot on the thighs. This, we
believe, is a very rare fpecies.

PLATE

PLATE CCCCXXXII.

GRYLLUS CAMPESTRIS.

FIELD CRICKET.

HEMIPTERA.

GENERIC CHARACTER.

Head inflected, armed with jaws: feelers filiform: antennæ ufually fetaceous or filiform: wings four, deflected, convolute, the lower ones plaited: pofterior legs formed for leaping: claws double.

** *Section* Acheta.. *Antennæ fetaceous: feelers unequal: thorax rounded: tail with two briftles.*

SPECIFIC CHARACTER

AND

SYNONYMS.

Wings fhorter than the wing-cafes: body blackifh: ftyle linear.

GRYLLUS CAMPESTRIS: alis elytris brevioribus, corpore nigro: ftylo lineari. *Lin. Muf. Lud. Ulr.* 124.
     Scop. Ent. Carn. 319.
     Fabr. Sp. Inf. I. p. 355. *n.* 10.
Gryllus Campeftris Mouffeti. *Ray Inf.* 63.
     Schaeff. Elem. t. 66.
     ——— *icon. t.* 157. *f.* 2—4

            Though

Though the Field Cricket inhabits every country of Europe, it is observed to be more abundant in the southern parts than elsewhere. Its haunts are shady places not too much exposed to moisture. The noisy chirpings of this singularly formed little creature is oftentimes heard issuing from among the bushes, and underwood, on the skirts of forests; and in the fields, towards the approach of twilight, particularly when the weather is warm and the air serene; but the insect itself is very seldom seen. It is remarkably timid, and scarcely ever ventures from its lurking place among the bushes, till the darkness of the night emboldens it to ramble out in quest of food.—Its chirping noise does not continue all the year; it commences in May, and ceases about the end of autumn. We imagine it almost superfluous to add, that the Field Cricket possesses this faculty of emitting a chirping note in common with many insects both of this, and other analogous tribes.

LINNÆAN

LINNÆAN INDEX

TO

VOL. XII.

COLEOPTERA.

HEMIPTERA.

LEPI-

INDEX.

LEPIDOPTERA.

NEUROPTERA.

HYMENOPTERA.

Apis

INDEX.

———

DIPTERA.

———

APTERA.

ALPHABETICAL INDEX

TO

VOL. XII.

Fur

INDEX.

Printed by Law and Gilbert, St. John's Square, Clerkenwell.

THE

NATURAL HISTORY

OF

BRITISH INSECTS.

Printed by Law and Gilbert, St. John's Square, Clerkenwell.

THE
NATURAL HISTORY

OF

BRITISH INSECTS;

EXPLAINING THEM

IN THEIR SEVERAL STATES,

WITH THE PERIODS OF THEIR TRANSFORMATIONS,
THEIR FOOD, ŒCONOMY, &c.

TOGETHER WITH THE

HISTORY OF SUCH MINUTE INSECTS

AS REQUIRE INVESTIGATION BY THE MICROSCOPE.

THE WHOLE ILLUSTRATED BY

COLOURED FIGURES,

DESIGNED AND EXECUTED FROM LIVING SPECIMENS.

———

By E. DONOVAN.

———

VOL. XIII.

LONDON:

PRINTED FOR THE AUTHOR,
And for F. C. and J, RIVINGTON, Nº 62, ST. PAUL'S CHURCH-YARD.
MDCCCVIII.

THE

NATURAL HISTORY

OF

BRITISH INSECTS.

─────────

PLATE CCCCXXXIII.

PAPILIO APOLLO.

APOLLO BUTTERFLY.

GENERIC CHARACTER.

Antennæ terminating in a club: wings erect when at reſt. Fly by day.

SPECIFIC CHARACTER

AND

SYNONYMS.

Wings entire, white ſpotted with black : lower ones with four red ocellated ſpots above and ſix beneath.

VOL. XIII.   B   PAPILIO

Papilio Apollo : alis albis nigro maculatis: posterioribus supra ocellis quatuor, subtus sex basique rubris. *Linn.* *Fn. Suec.* 1032.—*It. gothl.* 230.

Papilio Apollo. *Fabr. Inf.* 2. 35. *n.* 417. *Haworth. Lep. Brit.* 1. *p.* 29.

━━━━━━━━━━

This large and very beautiful butterfly, is an inhabitant of various parts of Europe, and is found also in the more temperate parts of Siberia. We are induced to insert it among the British Papillones on the assurance of Mr. Haworth, that he was recently informed the species had been taken in Scotland *. It would afford us much pleasure could the particulars of its capture be submitted to our readers, but we have been unable to procure any further information on this subject; and presuming the fact at least to be correctly stated, we could not refrain commencing our new volume with the representation of such an interesting object.

Papilio Apollo is the offspring of a solitary sluggish larva, or caterpillar, of a black colour, covered with a soft and silky down. All the rings are marked on both sides with two red spots, which together constitute a longitudinal series along each side. Besides these spots, every ring or joint is marked nearer the middle of the back with three small lateral dots of blueish, disposed in a semilunar manner, and thus forming a longitudinal waved line on each side within the two rows of red spots. The anterior part of the head is furnished with tentacula, which the animal can advance or retract at pleasure; this is furcated when completely spread out. The Pupa is slightly folliculate, somewhat ovate, and of a blueish colour.

---

* *Haw. Lep. Prodr. Pref. p.* 29.

PLATE

434

1

PLATE CCCCXXXIV.

APIS PENNIPES.

PLUME LEGGED BEE.

HYMENOPTERA.

GENERIC CHARACTER.

Mouth horny, jaw and lip membranaceous at the tip : tongue in-flected : feelers four, unequal and filiform : antennæ fhort and fili-form in the males, in the females fubclavated : wings flat : fting of the females and neuters pungent and concealed in the abdomen.

SPECIFIC CHARACTER

AND

SYNONYMS.

Somewhat greyifh and pubefcent: middle legs tufted with long hairs.

APIS PENNIPES : pubefcens fubgrifea ; pedibus fecundariis elonga-tis crinito-pectinatis. *Lin. Nat. MS. in Syft. Nat.*

APIS RETUSA *mas.* Corpore *mafculo* nigro, hirfuto-fulvo ; ano ni-gricanti, pedibus intermediis elongatis, crinito-pec-tinatis. *Kirby Ap. Angl. v. 2. p. 296. n. 69.*

APIS PLUMIPES, hirfuta, pedum mediorum metatarfis fcopa atra, pof-tice pilis raris longis barbatis. *Schranck. Enum. Inf. Auftr. n. 804.*

APIS PLUMIPES. *Pallas Spicil. Zool. 9. p. 24. tab. I. fig. 14.*

B 2 APIS

APIS PILIPES. *Chriſtii Hymenopt. p.* 131. *tab.* 8. *fig.* 9. *mas.*
APIS HISPANICA. *Panz. Fn. Inſ. Germ. Init. n.* 55. *tab.* 6.

―――――

This curious kind of bee appears to be deſcribed by ſeveral writers
under the various names of *plumipes, pilipes,* and *pennipes,* in allu-
ſion to the remarkable tufts of long hairs upon the middle pair of legs,
which contribute in ſuch a ſtriking manner both to the beauty and ſin-
gular appearance of the ſpecies. Panzer deſcribes it under the more
local, and therefore more objectionable epithet of *hiſpanica.*

Mr. Kirby conſiders this as no other than the male ſex of the Linnæan ſpecies *retuſa,* and notwithſtanding the great diſſimilarity which
prevails between the two inſects ſuppoſed to be male and female,
there is reaſon to apprehend, according to the obſervations of Mr.
Kirby, that they may be really of the ſame ſpecies : the evidence in
favour of ſuch an opinion affords a ſtrong preſumption of the fact, although it ſtill remains to adduce in ſupport of it, the teſtimony of thoſe
circumſtances which we are taught to admit as the moſt convincing,
and unerring proof.—It has been remarked, that when this inſect
makes its firſt appearance in the ſpring, the down of the upper ſide of
the body, the vent excepted, is of a reddiſh yellow colour, and that later in the year the hair becomes more cinereous.

This inſect is very local ; it is ſometimes found in neſts or cells con-
ſtructed in old walls in pretty conſiderable numbers, but except in
ſuch ſituations is very ſcarce. The ſmalleſt figure denotes the natural
ſize.

PLATE

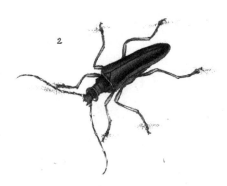

435

PLATE CCCCXXXV.

CERAMBYX SUTOR.

FIG. I.

COLEOPTERA.

GENERIC CHARACTER.

Antennæ fetaceous: feelers four: thorax fpinous or gibbous: wing-cafes linear.

SPECIFIC CHARACTER

AND

SYNONYMS.

Wings obtufe, deep black clouded with ferruginous; fcutel pale yellow: antennæ very long.

CERAMBYX SUTOR: elytris obtufis atris ferrugineo-fubnebulofis, fcutello luteo, antennis longiffimis. *Marfh Ent. Brit. T. I. p.* 329.—*Linn. Syft. Nat.* 628. 38.—*Fn. Suec.* 655.—*Gmel.* 1830. 68.
LAMIA SUTOR. *Fabr. Syft. Ent.* 172. 10.—*Spec. Inf. I.* 218. 15.
CERAMBYX ATOMARIUS. *De Geer.* 5. 65. 4.

———————

Cerambyx Sutor is not a common infect. The male differs from the female in being rather larger and having the antennæ three or four times the length of the body, the antennæ of the latter being much fhorter. The fpecies occur in woods.

FIG.

FIG. II.

CERAMBYX MERIDIANUS.

SPECIFIC CHARACTER

AND

SYNONYMS.

Black: wings fomewhat faftigiate, and with the tip of the abdomen teftaceous: breaft glofly.

Cerambyx Meridianus. *Linn. Syft. Nat.* 630. 47.—*Faun. Suec.* 648. *Gmel.* 1861. 47.

LEPTURA MERIDIANA: nigra, elytris fubfaftigiatis abdomineque apice teftaceis, pectore nitenti. *Marjh Ent. Brit. T. I. p.* 340. *n.* 1.

STENOCORUS MERIDIANUS. *Fab. Syft. Ent.* 178. 1.—*Spec. Inf.* I. 225. 1.—*Mant. I.* 143. 1.

———

The male of this fpecies is black, with the wing-cafes rufo-teftaceous, the female entirely black with a few yellow downy hairs,

PLATE

436

PLATE CCCCXXXVI.

SPHINX CRABRONIFORMIS.

LUNAR HORNET SPHINX.

LEPIDOPTERA.

GENERIC CHARACTER.

Antennæ thickeſt in the middle : tongue moſtly exſerted : feelers two, refleted : wings deflexed.

* Setion *Seſia.*

SPECIFIC CHARACTER

AND

SYNONYMS.

Head black : anterior margin of the thorax with a ſemilunar ſpot : abdomen yellow with two black bands.

Sphinx Crabroniformis. *Linn. Tranſ. Soc. v. 3. pl. 3. f.* 6—10.

━━━━━━

A ſpecies of the Seſia family, nearly allied to Sphinx apiformis, from which it is principally diſtinguiſhed by having the whole of the anterior margin of the thorax yellow inſtead of two ſpots of that co‐lour ; it is alſo rather ſmaller, and far more uncommon.

The

The larva is whitifh inclining to yellow, with fome brown dots; the pupa reddifh fufcous. Both fexes have two dark or blackifh bands at the bafe of the abdomen, but in one thofe bands are entire, in the other marked on each fide with a fubtriangular fpot of yellow. The larva lives in the trunks of willow trees, and appears in the winged ftate in July.

PLATE

PLATE CCCCXXXVII.

PAPILIO PRUNI.

BLACK HAIR-STREAK BUTTERFLY.

GENERIC CHARACTER.

Antennæ terminating in a club : wings erect when at reft. Fly by day.

SPECIFIC CHARACTER

AND

SYNONYMS.

Wings flightly tailed : above brown with a red fpot at the tip of the lower-ones : on the pofterior pair beneath a fulvous marginal band dotted with black.

PAPILIO PRUNI : alio fubcaudatis fupra fufcis : pofterioribus fubtus fafcia marginali fulva nigro punctata. *Linn. Fn. Suec.* 1071.—*Gmel. Linn. Syft. Nat.* 2341.
HESPERIA PRUNI. *Fab. Ent. Syft.* 3. 377. 70.

===

The larva of this fpecies is of a green colour with a pale lateral line ; the pupa brown with the head white.

This kind is found in the larva ftate on the cherry, bullace, and other trees of the Prunus genus, whence its name. It appears on the wing in July, and is not common.

VOL. XIII.   C   PLATE

1

PLATE CCCCXXXVIII.

APIS QUINQUEGUTTATA.

FIVE SPOT BEE.

HYMENOPTERA.

GENERIC CHARACTER.

Mouth horny, jaw and lip membranaceous at the tip: tongue inflected: feelers four, unequal and filiform: antennæ short and filiform in the males, in female subclavated: wings flat: sting of the females and neuters pungent and concealed in the abdomen.

SPECIFIC CHARACTER.

APIS QUINQUEGUTTATA. Black: second and third joint of the abdomen rufous: posterior ones black with five whitish dots.

SCOLIA QUINQUE-PUNCTATA. *Fabr. Ent. Syst.*

SAPYGA 5-PUNCTATA. *Latreille Inf.*

————

A scarce and very elegant species of the Melitta family. Our specimen was taken near Faversham in Kent.

The small figure resting on the leaf No. 1, represents the natural size; the upper figure is magnified.

c 2

PLATE

PLATE CCCCXXXIX.

PHALÆNA OLEAGINA.

GREEN BRINDLED DOT MOTH.

GENERIC CHARACTER.

Antennæ tapering from the bafe : wings in general deflected when at reft. Fly by night.

SPECIFIC CHARACTER

AND

SYNONYMS.

Wings green-brown with two white fpots, the anterior one pupilla-ted, pofterior largeft.

PHALÆNA OLEAGINA: alis viridibus fufco fubundatis: maculis duabus albis anteriore pupillata; pofteriore majore. *Wien. Schmetterl. p.* 59. *n.* 2.
GREEN BRINDLED DOT. *Haw. Lep. Brit. p.* 120. *n.* 70.

We accidentally met with an individual of this fpecies on the wing about twilight one fummer's evening * near Fifhguard in Pembroke-fhire, South Wales. It occurred among a number of other infects of the Noctua and Bombyx families by the fide of the low hedges which divide the fields and meadows at a fhort diftance from the town. This is the only fpecimen we have feen Britifh.

* In the month of July 1800.

The

The larva, as defcribed by Fabricius, is quadridentated: behind cinereous with black and red indentations, and the collar red dotted with black.

PLATE

PLATE CCCCXL.

SIREX BIMACULATUS.

BIMACULATED SAW-FLY.

HYMENOPTERA.

GENERIC CHARACTER.

Mouth with a thick, horny, truncated, fhort, denticulated mandible: feelers four, the pofterior part longer and thicker upwards: antennæ filiform, of more than twenty-four equal articulations: fting exferted, ferrated, and ftiff: abdomen feffile, terminating in a point: wings lanceolate and incumbent, the lower ones fhorter.

SPECIFIC CHARACTER.

Sirex bimaculatus. Abdomen yellow: a black fpot on the firft and three pofterior rings: thorax fubfufcous, with two black fpots.

———————

This is a curious fpecies, and differs apparently from any of the Sirex genus at prefent defcribed. The head and thorax is fufcous, the latter marked in the middle with two blackifh oblong fpots. The body is orange with black fpots, and the legs orange having the pofterior pair marked at the lower end of each principal joint with black. The legs are compreffed. Wings reddifh-brown. The figure in the plate is of the natural fize.

PLATE

441

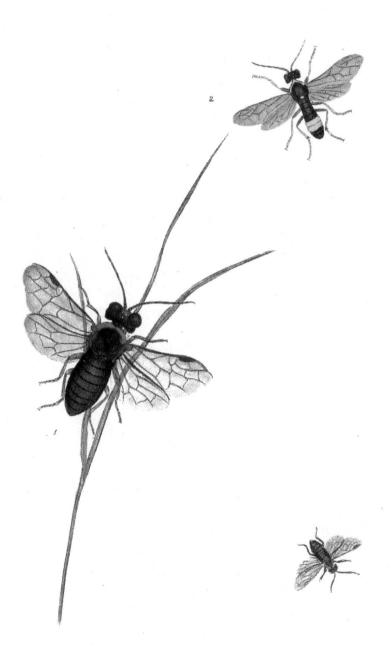

PLATE CCCCXLI.

FIG. I. I.

TENTHREDO COLLARIS.

COLLARED SAW FLY.

HYMONOPTERA.

GENERIC CHARACTER.

Mouth with a horny curved mandible, toothed within; the jaw ſtraight and obtuſe at the tip, the lip cylindrical and trifid: feelers four, unequal and filiform: wings tumid, the lower ones leſs: ſting com-poſed of two ferrated laminæ and almoſt ſecreted: ſtemmata three.

SPECIFIC CHARACTER.

TENTHREDO COLLARIS.   Black anterior margin of the thorax rufous.

TENTHREDO OPACA: antennis ſeptemnodiis atra thorace utrinque
macula apicis rufa.   *Fabr. Ent. Syſt. T. 2. p.* 120?

A ſcarce ſpecies ſhewn both in its natural ſize, and magnified.   It is found in gardens.

PLATE CCCCXLI.

FIG. II.

TENTHREDO SUCCINCTA.

BROAD YELLOW BANDED TENTHREDO.

SPECIFIC CHARACTER.

Black : margin of the thorax, fcutel, and two middle abdominal fegments yellow : legs yellow, thighs black.

———

Found on plants in May and June.

PLATE

PLATE CCCCXLII.

APIS LAGOPODA, *var.*

THICK LEGGED BEE.

HYMENOPTERA.

GENERIC CHARACTER.

Mouth horny, jaw and lip membranaceous at the tip: tongue inflected: feelers four, unequal and filiform: antennæ fhort and filiform in the males, in the females fubclavated: wings flat: fting of the females and neuters pungent, and concealed in the abdomen.

SPECIFIC CHARACTER

AND

SYNONYMS.

Grey: anterior legs dilated and ciliated: pofterior fhanks clavate: vent emarginate.

APIS LAGOPODA: grifea pedibus anticis dilato ciliatis, tibiis pofticis clavatis, ano emarginato. *Linn. Syft. Nat.* 2. 927. 27.—*Fn. Suec.* 1702.—*var.*
APIS LAGOPODA. *Panz. Fn. Suec.*

═══════════

The fingular dilation of the fhanks of the anterior legs of this infect form a very ftriking character of the natural family of bees, to which

D 2

the

the fpecies belongs. In the prefent infect the anterior legs are reddifh yellow; the dilation convex above, beneath convex, and deeply ciliated with hairs of the fame colour. It is a fcarce fpecies, and like its con-geries is fuppofed to live in the putrefcent wood of willow trees.

Fig. I. reprefents the upper furface of the infect in its natural fize; fig. 2. the underfide a little enlarged. Fig. 3, the under furface of the anterior leg. Fig. 4, the upper furface.

PLATE

PLATE CCCCXLIII.

PAPILIO RUBI.

GREEN HAIR-STREAK.

GENERIC CHARACTER.

Antennæ terminating in a club: wings erect when at reft. Fly by day.

SPECIFIC CHARACTER

AND

SYNONYMS.

Wings flightly tailed, above brown, beneath green.

PAPILIO RUBI: alis fubcaudatis fupra fufcis fubtus viridibus. *Linn.*
Fn. Suec. 1077. Gmel. Linn. Syft. p. 2352. 237.
HESPERIA RUBI. *Fabr. Ent. Syft. 523, 339.—Spec. Inf. 121.*
539.
Efper Schmet. t. 21. f. 2.
Ram. Gen. t. 18. f. 11. 12.
Schæff. Icon. t. 29. f. 5. 6.
Geoffr. Inf. p. 2. p. 64.

━━━━━━━━━

The green hair ftreak butterfly occurs in the larva ftate on the bramble in the month of July, and the perfect or winged infect appears in May.

The

The larva of this fpecies is of a green colour varied with yellow, andhas the he ad black: the pupa is pale brown.   The upper furface of the fly in both fexes is brown, and the lower a rich and elegant filky green.   Near the centre of the anterior wings on the upper fur-face is an obfcure paler fpot of an oblong form, and which is com-monly moft confpicuous in the male.   The two fexes may be in a great meafure diftinguifhed likewife by the number and fituation of the white dots on the lower furface of the pofterior wings, thefe in the fe-male conftituting an uniform feries or line extending entirely acrofs, while in the male they are interrupted, and though diftinct on the edges of the wing, are fcarcely perceptible towards the difk.   The fe-male is alfo larger than the male ; a circumftance not peculiar to this fpecies, but common to moft others of the infect tribe.

PLATE

PLATE CCCCXLIV.

TENTHREDO VIRIDIS.

GREEN SAW-FLY.

HYMENOPTERA.

GENERIC CHARACTER.

Mouth with a horny curved mandible, toothed within; the jaw ftraight and obtufe at the tip : lip cylindrical and trifid : feelers four, unequal, and filiform : wings tumid, the lower ones fmaller : fting compofed of two ferrated laminæ and almoft fecreted : ftemmata three.

SPECIFIC CHARACTER

AND

SYNONYMS.

Green : head and thorax, charactered with black : abdomen with black fpots.

TENTHREDO VIRIDIS : capite thoraceque fupra characteribus nigris. *Geoff. Inf.* 2. 271. *n.* 1.
Tenthredo viridis : antennis feptemnodiis, corpore viridi, abdomine fupra fufco. *Linn. Syft. Nat.* 2. 924. 27.—*Fn. Suec.* 1554.—*Fabr. Ent. Syft. T.* 2. *p.* 113. *n.* 33. *Sulz. Inf. tab.* 18. *fig.* 112.

―――――

The ground colour of this elegant infect varies from a pale yellow to bright green, and fometimes to brown, the markings of black lines

appear

appear to be pretty nearly the fame in all the varieties.   This infect is common in the fummer feafon, and is chiefly found on the Alder, on which it is fuppofed to feed.

The fmalleft figure in our plate denotes the natural fize.

PLATE

PLATE CCCCXLV.

FIG. I. I.

MUSCA MACULATA.

SPOTTED FLY.

DIPTERA.

GENERIC CHARACTER.

Mouth with a foft exferted flefhy probofcis and two equal lips: fucker furnished with briftles: feelers two, very fhort or fometimes none: antennæ generally fhort.

SPECIFIC CHARACTER

AND

SYNONYMS.

Cinereous: thorax lineated with black: abdomen fpotted with black, and marked at the tip with two black dots.

Musca maculata: antennis plumatis pilofa nigra, thorax nigro-
        lineato, abdomine atro-maculato, ano bipunctato.
        Linn. Syft. Nat. 12. 2. *p.* 990. *n.* 70.
        Scop. ent. carn. 870.
Musca maculata: cinerea, thorace abdomineque maculis nu-
        merofis atris. *Fabr. mant. Inf.* 2. *p.* 342. *n.* 8.

VOL. XIII.   E   This

This is a remarkably pretty fpecies, and appears to confiderable advantage before the lens of an opaque microfcope. The two fmall diftinct black dots on the extreme joint of the abdomen is a ftriking character of this interefting infect.

The fmalleft figure reprefents the natural fize; this fpecies is found on plants in Europe, and is rare.

FIG. II. II.

MUSCA SERICEA.

SILKY MUSCA.

SPECIFIC CHARACTER.

MUSCA SERICEA. Silky: head, thorax and fcutel fubteftaceous: firft three joints of the abdomen reddish-orange with a dorfal black ftripe; tip greyifh black.

A curious fpecies and of the fize denoted by the fmalleft figure (No. 2.) in the annexed plate. The whole furface of this infect except the wings has a delicate filky appearance: the thorax is marked in the middle with two diftant black lines, and each fide with two fhorter

lines

lines of the fame colour, forming altogether a remarkable character of this particular fpecies.

This kind appears to be rare, our fpecimen was taken in Kent in the neighbourhood of Faverfham.

E 2 PLATE

PLATE CCCCXLVI.

PAPILIO BRASSICÆ.

LARGE GARDEN WHITE BUTTERFLY.

GENERIC CHARACTER.

Antennæ terminating in a club : wings erect when at rest.   Fly by day

SPECIFIC CHARACTER

AND

SYNONYMS.

Wings rounded, entire, white : tip of the upper pair in the male black, and in the female marked with two black ſpots.

PAPILIO BRASSICÆ : alis anterioribus maculis duabus apicibusque
nigris, major.   *Linn. Fn. Suec.* 1035.—*Gmel.
Linn. Syſt. Nat.* 2259. *n.* 75.
Fabr. Ent. Syſt. 463. 110.—*Fab. Syſt. Ent.* 3.
186. 574.—*Spec. Inſ.* 2. 38. 161.
Roeſ. Inſ. 2. *t.* 4.
Eſper Schmet. t. 3. *f. I.*
Schaff. Icon. 40. *f.* 3. 4. *and* 140. *f.* 4. 5.
Geoffr. inſ. p. 2. *p.* 68. *n.* 40.
Wien. Schmetterl. p. 163. *D.* 2.

———————

In dry ſeaſons favourable to the growth and increaſe of theſe perni-
cious inſects, the larvæ become very injurious to our gardens, and
would

would be infinitely more fo were it not for the number of fmall birds which prey upon them, and thus lend their friendly aid to deftroy thofe unwelcome intruders. They feed for the moft part on cabbages, and fome other horticultural plants, which renders them more injurious to the kitchen garden than any other. The larva is of the number of thofe known in England by the trivial title of the grub, and in the perfect or winged ftate it is diftinguifhed by the lefs ambiguous epithet of Large Cabbage Butterfly.

The larva of this fpecies appears in fpring, and indeed throughout great part of the fummer, as there are two or more broods of them every year. The appearance of thefe Butterflies on the wing in a morning is confidered generally as an unerring indication that the weather will clear up, and the day eventually prove fine. This infect though common is certainly not uninterefting.

PLATE

447

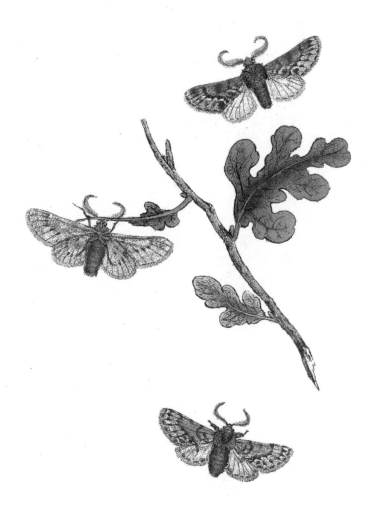

PLATE CCCCXLVII.

PHALÆNA URSULARIA.

THICK-HAIRED MOTH.

LEPIDOPTERA.

GENERIC CHARACTER.

Antennæ tapering from the bafe : wings in general deflected when at reft. Fly by night.

GEOMETRA.

SPECIFIC CHARACTER.

PHALÆNA URSULARIA. Thorax thickly clothed with hair : anterior wings pale brown, with undulated fufcous waves and dots : pofterior pair whitifh.

This moth poffeffes every character of a genuine Bombyx, and might be without fcruple referred to that fection, but for the authority of Mr. Drury who was fo fortunate as to rear it from the caterpillar; and which being of the looper kind decidedly proves it to be of the *Geometra* inftead of *Bombyx* family.

The antennæ in this infect as in the Bombyces is much pectinated, the thorax thick, and the body large and bulky. From its general habit it appertains therefore to the full bodied tribe of Geometræ, at the head of which we may rank the fpecies *Hirtaria*. This laft mentioned infect is fo clearly of the Bombyx family from its general appearance,

ance, that if its larva had not been diftinctly afcertained to be one of the loopers, we fhould have ftill confidered it of the Bombyx kind. Our prefent infect bears fome refemblance alfo to the fpecies Hirtaria but is fmaller.

Mr. Drury bred the two fexes of this extremely rare infect many years ago. Thefe fpecimens are ftill preferved in the cabinet formed by that eminent collector, and are the only examples of the kind we have ever feen; neither is the fpecies, which we believe to be perfectly new, in the poffeffion of any other collector within our knowledge.

Both fexes are reprefented in the annexed plate.

PLATE

PLATE CCCCXLVIII.

PAPILIO ADIPPE.

HIGH BROWN FRITILLARY.

LEPIDOPTERA.

GENERIC CHARACTER.

Antennæ clubbed at the end: wings erect when at reft: fly by day.

SPECIFIC CHARACTER

AND

SYNONYMS.

Wings indented, fulvous with black fpots; twenty three filver fpots on the pofterior pair beneath.

PAPILIO ADIPPE : alis dentatis fulvis nigro maculatis : fubtus ma-
culis 23 argenteis. *Linn. Fn. Suec.* 1066.—*Gmel.
Linn. Syft. Nat. p.* 2334. *n.* 212.—*Fabr. Ent.
Syft.* 3. 146. 448.—*Syft. Ent.* 517. 213.

The larva of this beautiful Butterfly is found on plants in May, and appears in the winged ftate in July. The larva is cinereous brown, with numerous rufous hairs and fpines, and a dorfal white ftripe, with a dark line along the middle, and the fides marked with a feries of white fpots. The pupa is fufcous with filvery dots.

F

This

This fpecies is fcarce, and occurs chiefly near the fkirts of woóds. The larva is faid to feed on the two fpecies of violet, viola odorata, and tricolor.

PLATE

449

PLATE CCCCXLIX.

LIBELLULA BIGUTTATA.

BIMACULATED DRAGON FLY.

NEUROPTERA.

Mouth armed with jaws, more than two in number; lip trifid: antennæ very thin, filiform, and ſhorter than the thorax : wings expanded : tail of the male furniſhed with a furcated proceſs.

SPECIFIC CHARACTER.

LIBELLULA BIGUTTATA. Abdomen depreſſed and narrow; the firſt joint marked in the middle with two ſmall yellowiſh ſpots.

A ſmall brood of this curious ſpecies of Dragon fly was diſcovered about eight years ago in a marſhy ground at Hampſtead, ſince which time they have entirely diſappeared. It is neither deſcribed by Linnæus or Fabricius, nor by any other writer we are acquainted with.

This new ſpecies is allied in its general aſpect to the Libellula depreſſa: the abdomen as in that inſect is flat, and riſing into a longitudinal ridge along the middle, but is conſiderably narrower in proportion, and this circumſtance is alone ſufficient to prove that it is of a different ſpecies. The pale yellowiſh ſpots on the firſt joint of the abdomen at baſe is alſo a ſtriking character : two of theſe ſpots are placed conguous to each other on the back, and two others appear one on each

F 2   ſide,

fide, but fo clofe to the lateral edge as to efcape attention unlefs in-
fpected clofely.   A fimilar lateral fpot is perceptible likewife a little
below thefe, neither of which are however fo confpicuous as the two
dorfal fpots firft mentioned.   The wings are tranfparent with a common oblong teftaceous ftigma at the coftal margin near the tip.

Libellula biguttata is reprefented in its natural fize in the annexed
plate.

PLATE

PLATE CCCCL.

PHALÆNA GRAMMICA.

FEATHERED FOOTMAN MOTH.

GENERIC CHARACTER.

Antennæ tapering from the bafe: wings in general deflected when at reft. Fly by night.

Bombyx.

SPECIFIC CHARACTER

AND

SYNONYMS.

Wings pale yellow: anterior pair yellow ftriated with black; lower ones with a black terminal band.

PHALÆNA GRAMMICA: alis luteis: primoribus flavis nigro ftriatis, pofterioribus fafcia terminali nigra. *Linn. Fn. Suec.* 1134.—*Amoen. acad. 5. t. 3. f.* 31.
Fabr. fp. inf. 2. p. 196. *n.* 113. *Roef. 4. t.* 21. *f. A. D. Geoffr. inf. p. 2. p.* 115. *n.* 17. *Schæff. icon. t.* 92. *f.* 2. *Ray Inf. p.* 169. *n.* 28. *and p.* 280. *n.* 13.

We introduce this elegant fpecies of the Moth tribe among the in-fects of Great Britain upon the moft fatisfactory authority, having

our-

ourfelves met with a living fpecimen of it in the Ifland of Anglefea fome few years ago. This occurred in the day time, in the month of September, under the fhade of a little clufter of ftones and bufhes near Manachty, the northern extremity of the ifland, and at no great dif-tance from the road to Gwyndy.

It is not entirely new as an Englifh infect, having been defcribed by Ray, but this is the only inftance within our recollection, fince the time of that writer, in which any naturalift is faid to have meet with it in our country. In Germany, and feveral others parts of Europe it is not very unfrequent.

The fpecimen difcovered by us is of the male fex, and is that re-prefented in the upper part of the annexed plate. The lower figure is of the female kind, which we have ventured to add, though taken from an exotic fpecimen in order to illuftrate the hiftory of this curi-ous infect, the two fexes of which differ fo materially that they might readily be miftaken for diftinct fpecies.

It rarely occurs to obfervation in the larva ftate: by fome it is fup-pofed to feed on the afh, while others affirm that its natural-food is the plaintain. The larva is brown with a white dorfal line, and rufous legs. The pupa ferruginous.

PLATE

451

PLATE CCCCLI.

CONOPS PETIOLATA.

PETIOLATED CONOPS.

DIPTERA.

GENERIC CHARACTER.

Mouth with a projecting geniculated probofcis: antennæ clubbed and pointed at the end.

SPECIFIC CHARACTER

- AND

SYNONYMS.

Antennæ black, with the club reddiſh: head yellow: abdomen petiolate.

CONOPS PETIOLATA: antennis nigris: clava rubra, capite flavo, abdomine petiolato. *Gmel. Linn. Syſt. Nat. p.* 2894.

An example of this very rare, and elegant infect occurs in Mr. Drury's cabinet of Britiſh infects now in our poffeffion: the particulars relative to its capture are however unknown to us. The ſpecies is evidently the *petiolata* of Gmelin, deſcribed by that writer, as it appears on the authority of Laxman as a native of Siberia. We have reaſon to apprehend that it is not figured in any work, and is only noticed as a ſpecies on the authority above mentioned.

The ſmalleſt figure denotes the natural ſize. The general colour of this infect is black with a hoary caſt, the petiole of the abdomen brownish brownifh red, and the club of the poifers yellow. Its wings are brown midway down from the coftal rib, and becomes hyaline towards the thinner margin. The legs are reddifh.

PLATE

PLATE CCCCLII.

PHALÆNA FASCIELLA.

BANDED TINEA MOTH.

LEPIDOPTERA.

GENERIC CHARACTER.

Antennæ tapering from the bafe : wings in general deflected when at reft. Fly by night.

Tinea.

SPECIFIC CHARACTER.

PHALÆNA FASCIELLA. Buff, with a broad greyifh figured band acrofs the middle : tip greyifh.

———————————————

An infect of fmall fize, but extremely elegant, and fingular in its appearance. We met with the fpecimen on the fide of Dinas Brân Hill, in the Vale of Llangollen, North Wales, in the month of Auguft, 1802.

This pretty acquifition is reprefented both in its natural fize and magnified.

PLATE CCCCLIII.

FIG. I.

PHALÆNA AUROSIGNATA.

SCARCE PURPLE GOLDEN Y MOTH.

LEPIDOPTERA.

GENERIC CHARACTER.

Antennæ taper from the bafe ; wings in general defle&ed when at reft : fly by night.

Noctua.

SPECIFIC CHARACTER.

PHALÆNA AUROSIGNATA. Anterior wings purplifh varied with fufcous: in the middle a vermicular golden charac-ter irregularly lobate at one extremity.

This is an extremely fcarce, and we believe, undefcribed fpecies. In its general appearance this curious infe&t is nearly allied to the No&tua interrogationis of Fabricius; it is as large as the Common Y Moth (*No&tua gamma*) : the anterior wings are tinged with purple, and the flexuous mark in the middle of each golden. The habits of this fpecies are unknown.

G 2

FIG.

FIG. II.

PHALÆNA LEUCONOTA.

WHITE-BACKED MOTH.

SPECIFIC CHARACTER.

PHALÆNA LEUCONOTA. Fufcous : thorax, back part of the anterior wings, and tranfverfe band white : abdomen and
pofterior wings whitifh.

———————

A fingle fpecimen of this elegant and ftrikingly fingular Phalæna,
occurs in the cabinet of the late Mr. Drury ; the figure in the annexed
plate reprefents it in its natural fize, and it is the only example of the
fpecies we have feen.

PLATE

454

PLATE CCCCLIV,

PAPILIO CRATÆGI.

BLACK VEINED WHITE BUTTERFLY.

GENERIC CHARACTER.

Antennæ terminating in a club: wings erect when at rest. Fly by day.

SPECIFIC CHARACTER

AND

SYNONYMS.

Wings white, entire; and veined with white.

PAPILIO CRATÆGI: alis albis: venis nigris. *Linn. Fn. Suec.*
1034.—*G l Li . yst. nat. p. 2257. n. 72.—*
Fabr. Ent. Syst. T. 3. . 82. 563.
Aldrov. inf. 246. f. 9.
Reaum. Inf. ?. t. 2. f. 9. 10.
Roef .f. I. p. 2. t. 3.
Frisch inf. 5. p. 16. t. 5.
Degeer inf. I. t. 14. fig. 19, 20.
Wien. Schmetterl. d. 163. D.
Schæff ic. t. 140. f. 2. 3.
Efper. pap. I. t. 2. j. 3.

━━━━━━━

Papilio Cratægi is one of the rarest species of the *Danai candidi,* or white tribe of butterflies found in Britain. It is a delicate, and by

no

no means inelegant infect, though altogether plain in its appearance; and may be readily diftinguithed from the other analogous fpecies, by the nerves both of the upper and lower wings being black, while the ground colour is white.

The larva feeds chiefly on the pear tree, or goofeberry, and is found in fpring. The perfect infect occurs in fummer, and is fometimes obferved in gardens hovering about fruit trees, the nectareous juices of which afford it fuftenance.

PLATE

1

PLATE CCCCLV.

VESPA SEXCINCTA.

SIX BELLED WASP.

HYMENOPTERA.

GENERIC CHARACTER.

Mouth horny, with a compreffed jaw : feelers four, unequal and filiform : autennæ filifiorm, the firft joint longeft and cylindrical : eyes lunar : body glabrous : fting pungent, and concealed within the abdomen : upper wings folded in both fexes.

SPECIFIC CHARACTER

AND

SYNONYMS.

Thorax fpotted : abdomen with fix yellow bands, the firft inter-rupted.

VESPA SEXCINCTA: thorace maculato: abdomine fafciis fex flavis primo interruptis. *Fabr. fp. inf. I. p.* 470. *n.* 7. *Mant. Inf. I. p.* 295. *n.* 9.

Defcribed by Fabricius as a native of Germany : it is a very rare Britifh fpecies, and has not been figured by any author.

PLATE

PLATE CCCCLVI.

PHALÆNA COMMUNIFASCIATA.

SINGLE STREAK MOTH.

LEPIDOPTERA.

GENERIC CHARACTER.

Antennæ taper from the bafe : wings in general deflected when at reft.   Fly by night.

GEOMETRA.

SPECIFIC CHARACTER.

PHALÆNA UNIFASCIATA.   Cinereous-buff with a fingle common fufcous ftreak in the middle.

———————————

An infect met with many years ago by Mr. Drury.   The upper wings are pale cinereous buff colour, the lower ones of the fame teint but rather paler, and both are marked with a common fufcous ftreak extending entirely acrofs the middle.   A fingle row of dots form a line along the outer margin of all the wings, and another feries though much fainter traverfe the upper wings midway between the common band, and the outer margin.   It is extremely rare, if not unique.

The figure reprefents this curious infect in its natural fize.

VOL. XIII.   H   PLATE

PLATE CCCCLVII.

BLATTA MADERÆ.

MADEIRA COCK ROACH.

HEMIPTERA.

GENERIC CHARACTER.

Head inflected: antennæ fetaceous: feelers unequal, and filiform: wing-cafes and wings fmooth, and fomewhat coriaceous: thorax rather flat, orbicular, and margined: legs formed for running: abdomen terminated in four fpines or briftles.

SPECIFIC CHARACTER

AND

SYNONYMS.

Brown: thorax livid variegated with brown: wing-cafes pale livid, the extreme half marked with numerous tranfverfe brown lines.

BLATTA MADERÆ: fufca thorace elytrifque lividis fufco variegatis. *Fabr. Ent Syſt. T. 2. p. 6. 119. 2.*

━━━━━━━━━━

A large fpecies of the cock roach tribe which inhabits the ifland of Madeira, and from whence it has been lately introduced into this country with goods and merchandize as the common cock roach (*Blatta orientalis*) was originally from the eaftern parts of the world. The fpecies has not yet become common in England. Our fpecimen was taken by Mr. Stachbury.

H 2　　　　　　　　PLATE

PLATE CCCCLVIII.

PHALÆNA GRAMINIS.

ANTLER MOTH.

GENERIC CHARACTER.

Antennæ tapering from the bafe : wings in general deflected when at reft.   Fly by night.

Bombyx.

SPECIFIC CHARACTER

AND

SYNONYMS.

Wings brown, with a trifurcated whitifh line, and dot.

PHALÆNA GRAMINIS : alis grifeis : linea trifurca punctoque albidis.
        Linn. Fn. Suec. 1140.—*Act. Stockh.* 1742. *p.* 40.
        t. 2.
        Fab. Spec. Inf. 2. *p.* 204. *n.* 148.
        —— *Syft. Ent.* 2. 586. 106.
        Frifch inf. 10. *t.* 21.
Noctua tricuspis *Hüb. Schmet.* 3. *t.* 60. *fig. I.*

This insect obtained the name of " *Antler Moth*" among the old collectors, in allufion to the trifurcated whitifh mark on the anterior wings, which bears fome refemblance to the antler of a ftag.   In this country the fpecies is very uncommon; in fome other parts of Europe on the contrary it is exceedingly abundant, and as it feeds on
                                                grafs

grafs is exceffively deftructive. The ravages committed in the fpace of a fingle year by this Infect in the Swedifh paftures has been efti-mated at a clear lofs of a hundred thoufand ducats.

The larva is fmooth, and dufky, with a dorfal yellow ftripe, and another of the fame colour on the fides. It is found at the roots of grafs; the winged infect in woods.

PLATE

PLATE CCCCLIX.

PHALÆNA BIMACULANA.

BIMACULATED TORTRIX MOTH.

LEPIDOPTERA.

GENERIC CHARACTER.

Antennæ tapering from the bafe : wings in general deflected when at reft. Fly by night.

Tortrix

SPECIFIC CHARACTER.

PHALÆNA BIMACULANA. Wings deep grey varied with tranfverfe fufcous and teftaceous bands : two large fub-triangular whitifh fpots on the interior margin.

━━━━━━━━━━

An elegant and curious moth of the Tortrix family found near Faverfham in Kent. The fpecies is not apparently defcribed by any author, and the prefent is the only fpecimen of its kind we recollect to have feen.

The fmalleft figure reprefents the natural fize.

PLATE

PLATE CCCCLX.

PAPILIO QUERCUS.

PURPLE HAIR-STREAK BUTTERFLY.

GENERIC CHARACTER.

Antennæ terminating in a club : wings erect when at reft : fly by day.

SPECIFIC CHARACTER

AND

SYNONYMS.

Wings flightly tailed; blue above, beneath cinereous with a white ftreak, and double fulvous dot in the anal angle.

PAPILIO QUERCUS : alis fubcaudatis cærulefcentibus, fubtus cine- reis : ftriga alba punctoque ani gemino fulvo. *Linn. Fn. Suec.* 1072.—*Mus. Lud. Ulr.* 314.—*Gmel. Linn. Syft. Nat. I p.* 2341.

HESPERIA QUERCUS. *Fabr. fp. inf.* 2. *p.* 113. *n.* 527.—*Mant. inf.* 2. *p.* 69. *n.* 652.—*Ent. Syft.* 3. 278. 72.
Efper Schmet. t. 19. *f.* 2. *c, a.*
Albin Inf. t. 52. *b. c.*
Admiral Inf. t. 17.
Roem. Gen. t. 18. *f.* 10.

The Purple Hair Streak feeds on the Oak. The larva is fat, of a pale or rofy red colour, and marked with lines of green dots; the pupa gloffy, and ferruginous, with three dorfal lines of brown

VOL. XIII.   I   dots

dots. This fpecies is found in the larva ftate in June, the fly appears in July.

Both fexes of this Butterfly are of a blackifh brown colour, but the male is diftinguifhed by having a large cordated fpace of a rich blue colour in the difk of the anterior pair. This is an interefting fpecies, and not by any means common.

PLATE

PLATE CCCCLXI.

FIG. I.

PHALÆNA MEDIOPUNCTARIA.

MIDDLE-DOT MOTH.

LEPIDOPTERA.

GENERIC CHARACTER.

Antennæ taper from the bafe : wings deflected when at reft. Fly by night.

Geometra.

SPECIFIC CHARACTER.

PHALÆNA MEDIOPUNCTARIA. Wings pale ; anterior pair with two tranfverfe waved lines and central dot of fufcous : pofterior pair with a fingle waved fufcous line.

A pair of this fingular and decidedly marked Phalæna occurs in the collection of Mr. Drury, and thefe are the only examples of the fpecies we have feen in any cabinet. It appears to be entirely of a new kind.

I 2

FIG.

FIG. II.

PHALÆNA TRISTRIGARIA.

THREE-STREAK MOTH.

SPECIFIC CHARACTER.

Phalæna Tristrigaria. Anterior wings greyiſh, with three
ſmall black ſtreaks at the tip, and a ſubteſtaceous
band acroſs the middle : poſterior wings immaculate.

———————————

This is an intereſting ſpecies of that particular kind of moths called
the Carpets.　The general colour is faint reddiſh grey with a ſingle
irregular broad band of a ſomewhat teſtaceous hue acroſs the middle
of the anterior wings, and the ſpace at the baſe of the wing is of the
ſame colour rather paler.　Towards the apex are three ſhort black
ſtreaks, and a ſmall triangular ſpot which ſeems to conſtitute one of
the moſt eſſential charaċteriſtics of this ſpecies.　The lower wings are
pale brown, and immaculate.

We believe this Phalæna is not deſcribed by any author.

PLATE

PLATE CCCCLXII.

PHILANTHUS FLAVIPES.

YELLOW LEGGED PHILANTHUS.

Vefpa *Linn.*

HYMENOPTERA.

GENERIC CHARACTER.

Mouth horny, with a compreffed jaw : feelers four, unequal and filiform : antennæ filiform, the firft joint longer and cylindrical : eyes lunar : body glabrous : upper wings folded in each fex : fting pungent, and concealed in the abdomen.

* Lip compreffed, rounded and longer than the jaw. *Philanthus Fabr.*

SPECIFIC CHARACTER

AND

SYNONYMS.

Black : thorax fpotted : abdomen yellow, with the edges of the fegments and tail black.

PHILANTHUS FLAVIPES : niger thorace maculato, abdomine flavo : fegmentorum marginibus anoque nigris. *Fabr. Ent. Syft. T. 2. p.* 290. *n.* 7.—CRABRO FLA-VIPES. *Fabr. Mant. Inf. I.* 295. 8.

Defcribed by Fabricius as a native of Italy, where it appears to be a rare infeĉt ; it is likewife found in Germany and England, but is not common in either country.

PLATE

PLATE CCCCLXIII.

PHALÆNA QUADRIPUSTULATA.

FOUR SPOT HEATH MOTH.

LEPIDOPTERA.

GENERIC CHARACTER.

Antennæ taper from the bafe : wings in general defle&ted when at reft : fly by night.

SPECIFIC CHARACTER.

Phalæna quadripustulata. Wings fcalloped, greyifh gloffed with purple, and fpeckled: a fmall whitifh fpot in the middle of each wing, and a common irregular line of dots behind.

━━━━━━━━

An elegant and very rare fpecies of the Geometra family with feta-ceous antennæ. This we difcovered fome years ago upon a fandy plain clofe to the Severn fea in the county of Glamorganfhire, be-tween Newton and Kenfig.

This Infe&t was taken in the winged ftate the laft week in July, and was obferved to fly in the open day, juft fkimming the furface of the fcanty herbage on the fands. Its flight was amazingly rapid.

PLATE

PLATE CCCCLXIV.

APIS VESTALIS.

HYMENOPTERA.

GENERIC CHARACTER.

Mouth horny, jaw and lip membranaceous at the tip: tongue inflected: feelers four, unequal and filiform: antennæ fhort and filiform in the males, in the females fubclavated: wings flat: fting of the females and neuters pungent and concealed in the abdomen.

SPECIFIC CHARACTER

AND

SYNONYMS.

Black, hairy, tail white with black tip: thorax at the bafe yellow.

Apis Vestalis: atra, hirfuta, ano albo, apice nigro; thorace bafi
flavo. *Kirby Ap. Angl. v.* 2. *p.* 347. *n.* 95.
Geoffr. Hift. Inf. Par. 2. *p.* 419. *n.* 26.
Apis vestalis. *Fourcroy Ent. Par. n.* 26.

———

The upper figure in the annexed plate reprefents the male, the lower the female, of this curious fpecies ; and both are fhewn in their natural fize. The female is largeft, and may be furthermore diftinguifhed by having the fulvous band at the anterior part of the thorax as well as the white fafcia at the pofterior part of the abdomen broader than in the other fex. The white abdominal band in the female is fomewhat

VOL. XIII. K

fomewhat interrupted by the black fpace at the tail pointing upwards, in the male its breadth is uniform throughout.

Mr. Kirby obferves that Apis veftalis is one of the Bombinatrices, which like A. campeftris and Barbutella have no inftruments for carrying or preparing maffes of pollen. The pofterior tibiæ, of one fpecimen in his cabinet is covered from one end to the other with a thin coat of pale earth, mixed with particles of fand, which it is probable they employ in conftruƈting their nefts or cells. The fpecies was known to Geoffroy, who defcribes it with accuracy except that he does not notice the black extremity of the vent*. The fpecies is found on flowers in the fummer.

" * Cette grande efpéce eft noire. Le haut ou la bafe de fon corcelet a une bande de poils jaunes citrons. Les deux tiers fupericurs du ventre font noirs, enfuite il y a quelques poils jaunes, et fon extrémité eft blanche."

PLATE

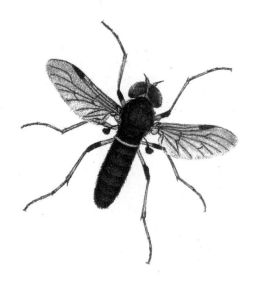

1

PLATE CCCCLXV.

MUSCA CINGULATA.

YELLOW GIRDLED MUSCA.

DIPTERA.

GENERIC CHARACTER.

Mouth with a foft exferted flefhy probofcis and two equal lips : fucker furnished with briftles : feelers two, very fhort or fometimes none : antennæ generally fhort.

SPECIFIC CHARACTER.

Musca cingulata. Deep velvetty black with a fingle yellow zone on the firft joint of the abdomen : legs yellowifh, extreme half of the thighs black.

Taken, though not in abundance, in the month of July, on the hedges near the road-side about Nutfield in Surrey. The fmalleft figure, No. I. exhibits the natural fize.

K 2

PLATE

PLATE CCCCLXVI.

PAPILIO PHLÆAS.

COMMON COPPER BUTTERFLY.

GENERIC CHARACTER.

Antennæ terminating in a club: wings erect when at reft. Fly by day.

SPECIFIC CHARACTER

AND

SYNONYMS.

Wings fub-entire, coppery fulvous fpotted with black, beneath hoary.

Papilio Phlæas: alis fubintegris fulvis nigro punctatis fubtus ca-
nefcentibus. *Linn. Fn. Suec.* 1078.—*Gmel. Linn.
Syft. Nat. I.* 2258.
Hesperia Phlæas. *Fabr. Ent. Syft.* 311. 178.—*Spec. Inf.*
126. 570.
Petiv. Muf. 24. *n.* 317.
Raj. Inf. p. 125. *n.* 20.
Roef. inf. 3. *t.* 45. *f.* 5. 6.
Geoffr. inf. p. 2. *p.* 65. *n.* 36.
Schœff. Icon. 143. 3. 4.

This fplendid little fpecies of Butterfly, is one of the moft familiar kinds, being very common in almoft every field and meadow, and de-
lighting

lighting in funny fituations among the hedges on the road fides, or the moft frequented foot paths, where it cannot eafily efcape the attention of the moft cafual obferver.

The general colour on the upper furface is fulvous richly gloffed with a metallic luftre, and finely relieved with a great number of black fpots, which contribute to render its appearance when expofed to the vigorous rays of funfhine equally brilliant and diverfified. Its larva and pupa feem to be unknown; in the winged ftate it appears very common from April till the end of Auguft.

PLATE

PLATE CCCCLXVII.

FIG. I. I.

MUSCA VIBRANS.

VIBRATORY FLY.

DIPTERA.

GENERIC CHARACTER.

Mouth with a soft exserted fleshy proboscis and two equal lips: sucker furnished with bristles: feelers two, very short, or sometimes none: antennæ generally short.

SPECIFIC CHARACTER

AND

SYNONYMS.

Wings hyaline with the tip black: head red.

Musca vibrans: alis hyalinis apice nigris, capite rubro. *Linn. Fn. Suec.* 1867.—*Degeer. Inf.* 6. *p.* 32. *n.* 11. *t.* 1. *f.* 19.

Found about trees, and distinguished in its flight by the brisk vibratory motion of its wings.

The figure resting on the leaf in the annexed plate denotes the natural size, that in the upper part of the plate its magnified appearance.

FIG.

FIG. II. II.

MUSCA QUADRISTRIGATA.

FOUR STREAKED FLY.

SPECIFIC CHARACTER.

MUSCA QUADRISTRIGATA. Brown : band furrounding the eyes, four ftreaks on the thorax, and fcutel yellow.

———————

A minute fpecies reprefented in its natural fize by the fmalleft figure in the lower part of the plate. Its magnified appearance is fhewn below.

PLATE

PLATE CCCCLXVIII.

FIG. I.

VESPA FLAVICINCTA.

YELLOW BANDED WASP.

HYMENOPTERA.

GENERIC CHARACTER.

Mouth horny, jaws compreffed: feelers four, unequal: antennæ filiform, with the firſt joint long and cylindrical: eyes lunate: body glabrous: fting pungent, and concealed: upper wings folded in both ſexes.

SPECIFIC CHARACTER.

VESPA FLAVICINCTA. Antennæ yellow, the extreme half black: head and thorax black, with yellow ſcutel: abdomen black banded with yellow.

———————

A ſpecimen of this inſect occurs in the cabinet of Mr. Drury, without any information relative to its capture. It is a ſpecies of the middle ſize as appears from the figure in the upper part of our plate, which exhibits the natural ſize. This inſect is not noticed in the works of any writer that we are acquainted with.

FIG. II.

AMMOPHILA ARENARIA.

LONG WINGED SAND-WASP.

GENERIC CHARACTER.

Snout conic, inflected, concealing a bifid retractile tubular tongue: jaws forcipated, and three toothed at the tip : antennæ filiform in each fex, and confifting of about fourteen joints: eyes oval : wings flat: fting pungent and concealed in the abdomen.

SPECIFIC CHARACTER

AND

SYNONYMS.

Black : petiole of a fingle articulation, and with the firft three joints rufous : wings length of the body.

SPHEX ARENARIA : hirta nigra abdominis petiolo uniarticulato : fegmento fecundo tertioque rufis, alis longitudine corporis. *Fabr. Ent. Syft. T. 2. p. 199. n. 2. Linn, Tranf. 4. p. 206.*

———————

Diftinguifhed from Sphex fabulofa to which it feems at firft glance nearly allied by the fhortnefs of the petiole which connects the body to the thorax, that part being remarkably long, and confifting of two joints in the above mentioned fpecies. Like that Infect Ammophila arenaria, inhabits fandy places, and appears moft lively in the fun fhine.—The figure in the lower part of the annexed plate reprefents this infect in its natural fize.

LINNÆAN

LINNÆAN INDEX

VOL. XIII.

COLEOPTERA.

HEMIPTERA.

LEPIDOPTERA.

L 2   Phalæna

INDEX.

ALPHA-

ALPHABETICAL INDEX

TO

VOL. XIII.

INDEX.

ERRATUM. VOL. XIII.

Plate 455, Veſpa ſexcincta.
Line 3, for *Six-bellied Wasp*, read *Six-belted Wasp*.

Law and Gilbert, Printers, St. John's Square, Clerkenwell.

THE

NATURAL HISTORY

OF

BRITISH INSECTS.

Law and Gilbert, Printers, St. John's-Square, London.

THE

NATURAL HISTORY

OF

BRITISH INSECTS;

EXPLAINING THEM

IN THEIR SEVERAL STATES,

WITH THE PERIODS OF THEIR TRANSFORMATIONS,
THEIR FOOD, ŒCONOMY, &c.

TOGETHER WITH THE

HISTORY OF SUCH MINUTE INSECTS

AS REQUIRE INVESTIGATION BY THE MICROSCOPE.

THE WHOLE ILLUSTRATED BY

COLOURED FIGURES,

DESIGNED AND EXECUTED FROM LIVING SPECIMENS.

By E. DONOVAN.

VOL. XIV.

LONDON:

PRINTED FOR THE AUTHOR,
And for F. C. and J. RIVINGTON, N° 62, ST. PAUL'S CHURCH-YARD.
MDCCCX.

469

THE

NATURAL HISTORY

OF

BRITISH INSECTS.

PLATE CCCCLXIX.

SPHINX DRURÆI.

DRURY'S HAWK MOTH.

LEPIDOPTERA.

GENERIC CHARACTER.

Antennæ somewhat prism-form, and tapering at each end: tongue generally exserted: feelers two, reflected; wings deflected.

SPECIFIC CHARACTER.

SPHINX DRURÆI. Wings entire: anterior pair grey and testaceous clouded, with distinct fuscous blotch in the middle: anterior wings red, with three denticulate black bands: abdomen red, with black belts.

SPHINX CONVOLVULI, *var.* POTATOE HAWK MOTH. *Smith's Inf. Georg. V.* 1. *p.* 32.

SPHINX CONVOLVULI, *varietas.* *Drury, V.* 1. *pl.* 25. *fig.* 4 ?

VOL. XIV.   B   A more

- 281 -

PLATE CCCCLXIX.

A more beautiful infect than that before us has never been introduced to the attention of our readers, either as an exotic species, or a native of this country; but, with what propriety we have ventured to confider it fpecifically diftinct from the Sphinx Convolvuli, to which it is fo clofely allied, or how far we may be authorized, from the occurrence of a fingle example in a living ftate in Britain, to admit it as an inhabitant, we are difpofed to fubmit to the decifion of others, after relating the circumftances which induce us to include it in the prefent work.

In a former volume our fubfcribers pöffefs a figure and defcription of another very interefting fpecies of the fame tribe, the Sphinx Caro- lina; an infect fufficiently known as a Linnæan fpecies, and as a na- tive of North America, but which was inferted as a Britifh infect on the authority of the late Mr. Drury, who received the individual fpe- cimen defcribed in a living ftate. It will be found, on reference to the memorandum in the hand-writing of Mr. Drury annexed thereto, that the information it conveys relates to two fpecies of the Sphinx tribe, the one we then defcribed, and another, which latter is the infect now under confideration. The memorandum ftates, that thefe two infects were brought to Mr. Drury alive, one about the year 1776, the other in 1788. Whether the fpecies Carolina, or the pre- fent, was difcovered firft, cannot be at this time afcertained: it is only evident that both were taken within the interval of the above-mentioned periods.

The difcovery of a folitary fpecimen of any infect in this country, which is clearly authenticated to be indigenous to extra European climates, is not altogether fufficient in our mind to countenance its introduction into the Britifh Fauna; yet there are circumftances, under which it would be improper to omit the mention of fuch extra- ordinary acquifitions; and this idea applies, in an immediate degree, to the difcovery of the prefent very elegant fpecies in a ftate of nature in Britain. We are neverthelefs inclined to regard it as an accidental occurrence .only, and conceive it incumbent to obferve, as in the inftance of Sphinx Carolina, that there appears to us every reafon

for

PLATE CCCCLXIX.

3

for believing it muſt have been originally imported in the egg, or larva ſtate, among ſome articles of American produce, though from this introduction it is not to be denied that the ſpecies may have become naturalized in this country. There does not appear any evidence ſo poſitive as to demonſtrate the fact, yet we ſuſpect this inſect, as a ſuppoſed variety of Sphinx Convolvuli, muſt have been long known among collectors as a native of Britain, under the denomination of the " Red Underwing Convolvuli ;" and, if we miſtake not, under that of the " Yorkſhire Convolvuli" alſo. We believe theſe names have been applied to the preſent inſect.

The ſimilarity that prevails in the general appearance of this inſect, and the Sphinx Convolvuli, deſerves particularly to be conſidered, in order to determine whether the latter be really a diſtinct ſpecies, or only a variety.

In the firſt place, it is to be obſerved, that the deſcriptions which Linnæus, and other early writers, afford us, are taken from ſpecimens of the Sphinx Convolvuli met with excluſively in Europe : thoſe writers did not conſider the ſpecies as extra European, much leſs as a native of the tranſatlantic regions, and their deſcriptions will be found to accord with that particular kind of Sphinx which is known in England by the name of Convolvuli, or Bind-Weed Hawk Moth.

Some time after the work of Linnæus appeared, our countryman Drury publiſhed the firſt volume of his exotic inſects, the twenty-fourth plate of which includes the figure of a Sphinx, whoſe external aſpect ſeemed, in his opinion, to correſpond with the European Convolvuli : the hues and marking of the upper wings were ſomewhat ſimilar, but in this the colour of the lower wings, which in the European inſect are greyiſh white, were red, a difference which the author of that work imagined might be produced from the effect of climate, the ſpecimen being from St. Chriſtopher's ; and under this perſuaſion, after ſpeaking of it as an inſect which he could not find deſcribed, he calls it in his index Sphinx Convolvuli *varietas*.

This

This induced later entomologifts, and among the reft Fabricius, to believe there muft be two varieties of the Sphinx Convolvuli, namely, the European kind with grey pofterior wings, and the American with red pofterior wings; for this, though not directly ftated, muft be implied, as he refers to the plate of Drury's work, before noticed among his fynonyms of the fpecies Convolvuli *. This latter infect was alfo, on fome popular report, confidered as a native of Britain, an idea we fufpect to have originated from its being underftood that an infect of the Sphinx family, correfponding with S. Convolvuli, but having red inftead of grey pofterior wings, had been once taken in England, and was preferved in the Englifh cabinet of Mr. Drury. Such we believe to be the origin of the report, though we cannot abfolutely trace it to this fource. Should this conjecture be well founded, we may add that the infect, figured and defcribed by Mr. Drury in his work, muft have been confidered different from the prefent fpecies by that author: it is very evident he did not admit them to be the fame; but whether the attention he had beftowed upon them was fufficient to enable him to determine this point with accuracy, we fhall not pretend to decide. Since the difperfion of his collection of exotic fphinges, it is perhaps impoffible to difcover the genuine infect intended by his Convolvuli *var.* His figure and defcription is not altogether fo definitive as we could wifh; and in the general information fubjoined thereto, he merely fays, " I received it from **St. Chriftopher's.** I cannot find it any where defcribed;" and after this, in the index, he names it " Convolvuli *varietas. Linn. p.* 798. *n.* 6." In his manufcript notes, at this time in our poffeffion, there is a further memorandum on the fame fubject, and which, though not material, may be repeated. It occurs in the following words: " Convolvuli vâr. **St. Kitt's. Mr. Kearton,** 1765." *Vid. Illuft. Vol. I. pl.* 25. *fig.* 4. In the manufcript note annexed to our prefent infect, Mr. Drury expreffes a different opinion of the latter; for this, he obferves, " is not the fame as S. Convolvuli;"

---

* The reference in Species Infectorum is to plate 25. fig. 1. which latter is an error; it is intended for figure 4. The fame error has been followed by Gmelin in his Linn. Syft. Nat. but this is corrected in the more recent works of Fabricius.

from

PLATE CCCCLXIX.   5

from which it is to be inferred, that he confidered the firft of thefe infects as only a variety of S. Convolvuli, and the latter as a diftinct fpecies. We fhall not, however, adduce this as a pofitive teftimony that they were in reality different: indeed we fufpect the contrary; but on a fubject fo ambiguous, we conceive it candid to ftate the ideas of Mr. Drury, as well as the opinion we ourfelves entertain.

Since the production of the work to which we laft adverted, Mr. Abbot, an affiduous entomological collector in the province of New Georgia, North America, furnifhed feveral of the Englifh cabinets with fpecimens of the infects of that particular country where he refided, and among the reft with fome few examples of the individual kind of Sphinx to which our attention is now directed. A feries of drawings by Mr. Abbot, explanatory of the various changes of a felect number of the infects of that part of the globe, were likewife tranfmitted to England about the fame period, one of which exhibited the transformation of this very fpecies. Thefe drawings afterwards paffing into the hands of the London bookfellers, were engraven and publifhed under the title of Abbot's Infects of Georgia, with obfervations by Dr. Smith.

Thus it appears, that of the two figures confidered as reprefentations of our infect, one only is certain, and that is the figure included in the laft mentioned publication. The latter we are affured of, not only from an attentive infpection of the original drawings *, but alfo from the individual example delineated in that work, and which differs in no refpect from the infect now before us. This we mention in order to fhew that our comparifons are deduced with a fufficient degree of certainty.

---

* Thefe original drawings were, in the firft inftance, configned from Georgia by Mr. Abbot to Mr. G. Humphreys, in London, and remained in the poffeffion of the latter fome time. They were executed by Mr. Abbot on coarfe wire-marked paper, and were, more or lefs, difcoloured and ftained with fea-water, an injury fuftained in the paffage between America and England. With the exception of this circumftance, we have no reafon to diftruft their general accuracy, and that exhibiting the transformations of our prefent infect had in particular efcaped without any material damage.

From

From the remarks of Dr. Smith on this particular fubject, it is obvious he confidered it only as a variety of the European kind of Sphinx Convolvuli. " We cannot difcover," fays this author, " any material diftinction between this and the moth which feeds on plants of the fame genus * in Europe, and is often feen fluttering about in towns and houfes, making as much noife as a bat, or fmall bird, for both which it is often taken by the vulgar. The reddifh tinge on the under-wings of the American one, is the only difference we can find, and is furely not fufficient to make that kind any more than a variety, as Mr. Drury fuppofes it. Fabricius does not even diftinguifh it as fuch. Mr. (now Dr.) Latham informs us, this variety has been found in England."

Before we offer any obfervations likely to difcountenance the perfuafion of this refpectable writer, it will not be amifs to ftate, that it appears to have been uniformly the idea of every entomologift, as well as Dr. Smith, with the exception of Mr. Drury, that our infect is only a variety of Sphinx Convolvuli. Mr. Drury remarks, in the manufcript note above adverted to, that they are certainly different, and that this difference is manifeftly difcernible. But while we rely on the defcription which Linnæus affords of the fpecies, it is perfectly confiftent to maintain the contrary opinion ; and it was hence depending on the Linnæan character, that in our defcription of Sphinx Convolvuli, we were inclined to fpeak of the prefent infect as a variety of the former, rather than as a new fpecies. In adverting to the paffage in which this fuppofed variety was mentioned, it will be however perceived, that we entertained, at that time, no inconfiderable degree of diftruft as to the propriety of fuch an opinion, for it was then obferved, that " it has all the characteriftic marks of Sphinx Convolvuli, or we fhould hefitate to admit it as the fame fpecies." Such were the fcruples at that time prevalent in our mind : we were unwilling to oppofe the authority of Linnæus, or we fhould have then conftituted it a diftinct fpecies. Subfequent obfervations have tended only to ftrengthen the propriety of this fuggeftion, and to convince us, the Linnæan cha-

---

* Feeds in America on the fweet potatoe, Convolvulus Batatus.

racter

PLATE CCCCLXIX.   7

racter of the species Convolvuli is too indefinite to form any precise criterion of the species.

On the latter topic we wish to speak more fully in explanation. There is nothing, we would obferve, laid down in the Linnæan character to prove the two above-mentioned infects diftinct; but, on the contrary, every character is calculated to confirm it. Linnæus had not, in all probability, feen this fuppofed variety : his fpecifical definition was apparently drawn from examples of the European Convolvuli ; and he was doubtlefs not aware that the character he affigned thereto was fo far inapplicable as to apply to two diftinct infects ; thefe according in every character with the fpecifical diftinction he propofes, though in other refpects they are remote from each other. Hence it is obvious, that our prefent infect may really, according to that character, be the Sphinx Convolvuli, or Convolvuli *vár.* of Linnæus, though as a fpecies it may be ftill diffimilar. The accuracy of this obfervation will be more amply demonftrated from the following comparifon of the two infects, at prefent under confideration, with the fpecific character which Linnæus affords of the Sphinx Convolvuli.

Linnæus, in the earlier editions of his Syftema Natura, thus defines the laft mentioned fpecies :—" Alis integris pofticis albo fafciatis margine poftico albo punctatis, abdomine rubro cingulis atris." According to which, the two infects before us would be at once diftinguifhed as fpecifically diftinct, the bands on the pofterior wings being red in one, and white, or at leaft greyifh white, in the other.

This defcription occurs in the tenth edition of the Syftema Natura, and it is poffible, though it appears otherwife expreffed in the later editions of that work, that Linnæus ftill intended to preferve the fame interpretation : it would be uncandid to conclude the contrary, though his words may bear a different acceptation, becaufe he does not himfelf contradict this fuppofition. It appears, however, confining our attention folely to the defcription given of the fpecies in the twelfth edition of that work, and in the fubfequent editions publifhed by Gmelin, that the two kinds may be ftill confounded, the colour of the paler bands forming,

forming, according to thofe defcriptions, no criterion of the fpecies.
In the laft mentioned work, the S. Convolvuli is thus defcribed :—
" Alis integris, pofticis nigro-fafciatis margine poftico albo punctatis,
abdomine rubro cingulis atris." And this defcription will be found
applicable to either of the infects before us: in both the wings are
entire, the pofterior pair barred with black, the hinder margin
dotted with white, and the abdomen red, with belts of black.

The Fabrician character of S. Convolvuli :—(" Alis integris nebulofis, pofticis fubfafciatis abdomine cingulis rubris atris albifque."
Syft. Ent. 544.) will alfo agree very nearly with either : the wings in
both are entire, and clouded: in both the pofterior wings are barred,
though flightly in Convolvuli, and confpicuoufly in the other, and in
each the abdomen is belted with black and red, though in Convolvuli
every fegment is marked at the bafe with a band of white, no trace of
which appears in the other.

From the above it will be inferred, that the defcription which the
lateft work of Linnæus offers will correfpond with both the infects in
queftion, and that of Fabricius will alfo accord in almoft every effential
particular ; notwithftanding which, we are perfuaded, for the following
reafons, they ought to be confidered as diftinct :—

1. The Sphinx Convolvuli, fo far as we have been enabled to compare the two kinds, is rather larger: this difference, we admit, may
arife from the influence of climate, or any other adventitious caufe.

2. There is a flight difference in the contour, the curvature in the
floping margin of the wings being moft diffufe in S. Convolvuli.

3. The anterior wings in both are clouded and greyifh, but in our
prefent infect the grey is finely varied with ochraceous hues ; and
there is, befides, in the middle of the wings of the latter, a perfectly
characteriftic fufcous blotch, margined behind with an irregular greyifh
fubcatenated band, neither of which appear in the wings of S. Convolvuli.

4. In

PLATE CCCCLXIX.

9

4. In both kinds the anterior wings are tranfverfely barred, or lineated with a number of indented dark ftreaks, but in the form of thofe the moft obvious difference prevails. Thefe lines are moft numerous in S. Convolvuli, and are in that infect fo deeply indented as to exhibit a lozenge-form zic-zac, the arches (if the expreffion be allowable) being greatly elongated, and extending into an acute falient point. In our prefent infect, the correfponding lines are difpofed acrofs the anterior wings, in a fimilar manner; but thefe, befides being lefs confiderable in number, are neither zic-zac, nor pointed, for though indented, the angles are almoft uniformly rounded, fo as to affume a fcalloped inftead of pointed arch-like appearance.

5. Another difference fubfifts in the under wings, and which, as well as that of the upper wings, is confiderable. In S. Convolvuli the prevailing colour is grey, in the prefent fine rofe-colour; in S. Convolvuli the black bands are four in number, in the prefent only three. The two middle bands in fome examples of S. Convolvuli are indeed confluent, but in no inftance whatever have wef een thofe bands fo clofely united as to conftitute only a fingle apparent band; while in our prefent infect, the middle of the wings are traverfed by a fingle band only, and that of a black colour, far more intenfe than we have ever obferved in the bands of S. Convolvuli.

6. The larva or caterpillar of Sphinx Convolvuli is of a fine green colour, with a fingle narrow darker green line along the back; each of the fegments alfo are marked on the fides with an oblique whitifh yellow line, edged above with dufky or blackifh; and four dufky fpots, two of which are placed adjacent to the anterior part of the dorfal line, and the others are on each fide contiguous to the fpiracles. This is the laft appearance it affumes before it paffes into the pupa form; in the ftate previous to this last appearance, its colour is brown, with the fides ochraceous. The larva of our prefent infect we have not feen, but from the drawing made by Mr. Abbot, and which we have attentively compared with the former, there can remain no doubt of its being altogether a diftinct fpecies. Thefe caterpillars, according to Abbot, are frequent in Georgia, though the moth is rare, and in the

former ftate its appearance muft be familiar, therefore, to this affiduous collector. The prevailing colour in this delineation is brown, with longitudinal ftripes of pale orange, rofy-white and yellow. Along the upper part of the back is ·a broad ftripe of faint orange, inclofing, on each joint, an oblong, or fomewhat fhuttle-form fpot of black, and which altogether exhibits a flightly interrupted or fubcatenated band : this is fucceeded beneath by a fufcous band of moderate breadth : a line ftill narrower, and of a delicate rofy-white, runs parallel to this lower edge of the fufcous band ; and beneath that the body is brown, with the exception of a yellow band, difpofed immediately under the feries of fpiracles. The laft mentioned band extends throughout the whole length, but is confluent on the anterior part of each fegment, and there becomes fo much produced and curved backwards as to appear deeply falcated. This is the laft fkin of the larva according to Abbot.

7. The difference in the pupa ftate is not confiderable : they are nearly of the fame form and colour; a fimilarity in this ftate is, however, obfervable in many infects of very different fpecies.

8. Neither is it conclufive, from the nature of their food, that they muft be fpecifically allied, as vaft numbers of very diffimilar infects are known to fubfift on plants of the fame kind.—The European S. Convolvuli feeds on the common bindweed Convolvulus Major, and the Georgian infect on the Convolvulus Batatas.

9. The time in which the Sphinx Convolvuli makes its firft appearance in the winged ftate, is about the middle of September. The larva of the other, Mr. Abbot informs us, went into the ground on the 20th of Auguft, and the fly came forth on the 11th of September: this was in Georgia ; but in Virginia, where he met with the fame fpecies, a larva of this kind buried itfelf on the 3d of October, and did not produce the fly till the 30th of May following.

We have thus endeavoured to ftate precifely every material circumftance, within our own knowledge, that could poffibly tend to deter-

mine

PLATE CCCCLXIX.   11

mine in what particulars the two above-mentioned infects accord or disagree. For the prolixity of our statement we may claim some indulgence, as it was deemed incumbent to shew, that we were not disposed, on very trivial grounds, to contradict an opinion so generally prevalent, as that of the present infect being a variety only of S. Convolvuli; an opinion that seems to have obtained an uniform ascendancy over the minds of entomologists in this country, and apparently of some on the continent also. Those infects, when examined with scrupulous attention, appear indeed to differ in so many essential respects, that it would seem impossible they could heretofore have been considered fully, or we apprehend it would not have remained for us to point out their differences. Upon the whole, therefore, we feel impressed with the propriety of considering them specifically distinct, though, at the same time, it must be acknowledged, at the first view, they might be casually admitted as varieties of each other.

C 2   PLATE

1

PLATE CCCCLXX.

SCARABÆUS GLOBOSUS.

GLOBOSE BEETLE.

COLEOPTERA.

GENERIC CHARACTER.

Antennæ clavate, the club lamellate; feelers four; the anterior ſhanks uſually denticulate.

SPECIFIC CHARACTER

AND

SYNONYMS.

SCARABÆUS GLOBOSUS. Gloſſy blackiſh: head granulated: wing-
cafes ſtriated.

ÆGIALIA GLOBOSA. *Latr. Gen. Cruſt. et Inſ.*—Aphodius. *Illig.*
Panzer.

━━━━━━━

A few years ago we diſcovered this curious inſect in ſome plenty, feeding, as it appeared, on the remains of certain marine vermes of the Meduſa tribe, thrown on the ſandy ſhore of Barmouth, in the great bay of Cardigan, North Wales. Before that period we have reaſon to conclude this inſect was unknown: it has been ſince deſcribed by Panzer and Latreille, both of whom mention it as an inhabitant of maritime marſhes. The fame inſect has been alſo taken ſince we obſerved

ferved it at Barmouth, in fimilar fituations, in other parts of Britain. Mr. Hooke met with it near Hull, and Mr. Leach, at Clonkelty, in Ireland. It therefore appears, upon the moft fatisfactory information, to be a local fpecies, and one confined to marfhy and fandy places in the vicinity of the fea.

P L A T E

PLATE CCCCLXXI.

FIG. I.

MUSCA MYSTACEA.

DIPTERA.

GENERIC CHARACTER.

Mouth with a foft exferted flefhy probofcis, and two equal lips: fucker furnifhed with briftles : feelers two, and fhort, or fometimes none : antennæ generally fhort.

SPECIFIC CHARACTER

AND

SYNONYMS.

Black : head, margin of the thorax and tip of the abdomen yellow.

Musca mystacea : nigra, thorace abdominisque apice flavis. *Linn. Fn. Suec.* 1793.
Syrphus myftaceus. *Fabr. Spec. Inf. T. 2. p.* 421. 175. 1.— *Schæff. Elem. t.* 131.—*Icon. t.* 10. *f.* 9.

Inhabits woods.

FIG.

FIG. II.

MUSCA MERIDIANA.

SPECIFIC CHARACTER.

Hairy, black: front golden: wings ferruginous at the bafe.

Musca meridiana: nigra, fronte aurea, alis bafi ferrugineis.
Linn. Fn. Suec. 1827.
Fabr. Sp. Inf. 2. *p.* 435. *n.* 3.
Mufca nigra, alis bafi ferrugineis. *Geoff. Inf.* 2. 495. 5.

———————

Common in woody places throughout moft parts of Europe.

PLATE

PLATE CCCCLXXII.

LIBELLULA CANCELLATA.

CANCELLATED DRAGON FLY.

NEUROPTERA.

GENERIC CHARACTER.

Mouth armed with jaws, more than two in number: lip trifid: antennæ very thin, filiform, and fhorter than the thorax: wings expanded: tail of the male furnifhed with a forked procefs.

SPECIFIC CHARACTER

AND

SYNONYMS.

Wings immaculate at the bafe: abdomen on the back and fides interrupted with yellow.

LIBELLULA CANCELLATA : alis bafi immaculatis, abdomine, dorfo lateribufque interrupte luteis. *Linn. Fn. Succ.* 1465.—*Syft. Nat.* 12. 544.—*Fab. Spec. Inf.* 1. *p.* 522. *n.* 15.—*Mant. Inf.* 1. *p.* 337. *n.* 15.

———

The defcription which the work of Linnæus affords of his Libellula Cancellata is remarkable for its brevity, and as he refers to no other authority, fome diftruft might arife as to the identity of the infect intended, were it not materially different from the other European fpecies; infomuch, indeed, that it cannot eafily, we fhould imagine, be

VOL. XIV.   **D**   confounded

confounded with any other : the defcription, though concife, is expreffive, and perfectly applicable to the infect before us. Fabricius fpeaks
of the fpecies only in the words of Linnæus.

There is a figure of one of the Libellulæ in the work of Sulzer,
given under the name of Cancellata, which nearly refembles the prefent
infect, but is fcarcely more than half its fize ; and this figure is repeated
from the fame plate in the work of Roemer, but neither is referred to
by Fabricius. The magnitude of L. Cancellata is not fpecified ; the
figure appears to be tolerably correct, fo far at leaft as to be underftood, and we fhould rather fufpect it to be a dwarf example of the
fame infect.

The " *Icones Inf. circa Ratifbon.*" of Schæffer, plate 137, fig. 1.
prefents another reprefentation, if we miftake not, of the fpecies
Cancellata, in fize approaching much nearer to the fpecimen delineated
in the annexed plate. It appears without any fpecific name, as ufual,
in that work. Thefe are the only figures we at prefent recollect, that,
in our opinion, are to be efteemed fynonymous.

This interefting infect is delineated from a fpecimen in the cabinet
of Mr. W. Leach, F. L. S.

P L A T E

PLATE CCCCLXXIII.

CIMEX NIGRO LINEATUS.

BLACK-LINED ELDER BUG.

HEMIPTERA.

GENERIC CHARACTER.

Snout inflected : antennæ longer than the thorax : wings four, folded acrofs, the upper pair coriaceous on the fuperior part : back flat : thorax margined : legs formed for running.

SPECIFIC CHARACTER

AND

SYNONYMS.

Red : thorax with five black lines, fcutel with three : abdomen yellow, with black dots.

CIMEX NIGRO-LINEATUS : ruber : thorace lineis quinque, fcutello tribus nigris, abdomine flavo : punctis nigris. *Fabr. Spec. Inf.* 2. *p.* 341. *n.* 15.—*Mant. Inf.* 2. *p.* 281. *n.* 17.—*Gmel. Linn. Syft. Nat.* 2131. *n.* 6.—*Schæff. elem. t.* 44. *f.* 1. *Icon. t.* 2. *f.* 3.

———

This beautiful infect is found in vaft abundance in the fouth of Europe, and infefts the flowers of the elder : in Britain the fpecies is very rare.

D 2   **PLATE**

1

PLATE CCCCLXXIV.

BEMBEX OCTO-PUNCTATA.

OCTO-PUNCTATED WASP.

HYMENOPTERA.

GENERIC CHARACTER.

Mouth horny, with arched and pointed jaws: tongue inflected and quinquefid: upper lip much advanced: feelers four, fhort, unequal, filiform: antennæ filiform, the first joint thrice the length of the others: eyes large, and occupying the whole fides of the head: body glabrous: fting pungent, and concealed in the abdomen.

SPECIFIC CHARACTER.

BEMBEX OCTO-PUNCTATA.   Greenifh, varied with bands and lines of black: two black dots on each of the firft four fegments of the abdomen.

The fmaller figure in the annexed plate denotes the natural fize of Bembex octo-punctata; the enlarged reprefentation being intended to exprefs its appearance before the lens of the opake microfcope.

We are not aware that any fpecies of the Bembex genus has been before defcribed or mentioned as a native of this country. The genus is rather limited in point of number, and is confined, with few exceptions, to extra European climates.   Bembex roftrata is the moft common of the European kinds, and is found in France, and other

parts

parts of the continent, in fome abundance, but has never, to our knowledge, occurred in England. Our prefent infect, and which is probably the only example of its kind hitherto difcovered in this country,
was taken by the late Mr. Drury, and is preferved in his cabinet now
in our poffeffion. Though extremely rare, it is not, however, to be
confidered as an unique infect, except as a Britifh fpecies, for we have
obferved two examples of the fame kind in the fplendid entomological
collection of our worthy friend, A. M'Leay, Efq. F. R. S.

P L A T E

475

PLATE CCCCLXXV.

SCOLOPENDRA HORTENSIS.

GARDEN CENTIPEDE.

APTERA.

GENERIC CHARACTER.

Antennæ fetaceous : feelers two, filiform, and united between the jaws : lip toothed and cleft : body long, depreffed, confifting of numerous tranfverfe fegments : legs numerous.

SPECIFIC CHARACTER.

SCOLOPENDRA HORTENSIS : fufcous : legs on each fide twenty-one.

━━━━━━━━

This Centipede appears to be of an undefcribed fpecies : it was difcovered, in fome abundance, by Mr. W. Leach, in the gardens at Exeter.

The natural fize of this infect is delineated in the lower part of the annexed plate ; and from this it will be obferved, that the fpecies is of the more diminutive kind. In its general afpect it bears a very ftrong analogy to the great venemous Centipede of the eaftern parts of the world, Scolopendra morfitans. The refemblance is indeed fo ftriking, that notwithftanding the difparity of fize, were it not for the prefent fpecies differing, in having a pair of legs more than that infect, we fhould not be inclined to think it fpecifically diftinct. The number of legs in the Scolopendra genus is admitted by entomological writers

writers as a criterion of the fpecies, and for this reafon it is fubmitted as a new infect.

The figure in the upper part of the plate exhibits its magnified appearance.

PLATE

476

PLATE CCCCLXXVI.

FIG. I. I.

ICHNEUMON LEUCORHÆUS.

WHITE-TAILED ICHNEUMON.

GENERIC CHARACTER.

Mouth with a ftraight horny membranaceous bifid jaw: the tip rounded and ciliated: mandibles curved, fharp; lip cylindrical, membranaceous at the tip, and emarginate; feelers four, unequal, and filiform, and feated in the middle of the lip: antennæ fetaceous, of more than thirty articulations: fting exferted, inclofed in a cylindrical fheath compofed of two valves, and not pungent.

SPECIFIC CHARACTER.

ICHNEUMON LEUCORHÆUS. Head and thorax black: body fub-globofe, and rufous, terminating in a black band, and yellowifh white tip.
Ichneumon octogefimus primus. *Schæff. Icon. pl.* 187. *fig.* 1. ?

The fmaller figure denotes the natural fize of this curious infect, the larger being confiderably magnified. The globofity of the abdomen is remarkable, but not peculiar to this fpecies: its legs are brown and black, and the antennæ rather longer than the wings. We have reafon to believe this a rare fpecies.

FIG. II. II.

ICHNEUMON COSTATOR.

YELLOW-MARGINED ICHNEUMON;

SPECIFIC CHARACTER.

ICHNEUMON COSTATOR. Head and thorax black: body black, with the furrounding margin, and edge of the fegments yellow.

————————

A minute fpecies, the natural fize of which is reprefented by the fmaller figure, No. I.

P L A T E

477

PLATE CCCCLXXVII.

FIG. I.

CARABUS SYCOPHANTA.

GENERIC CHARACTER.

Antennæ filiform : feelers generally fix, the laft joint obtufe and truncated : thorax flat and margined : wing-cafes marginate.

SPECIFIC CHARACTER

AND

SYNONYMS.

Winged : fhining violet : wing-cafes green-gold and ftriated.

CARABUS SYCOPHANTA : alatus violaceo-nitens : elytris ftriatis au-reis. *Fabr. Sp. Inf.* 1. *p.* 303. *n.* 25.—*Mant. Inf.* 1. *p.* 197. *n.* 34.

CARABUS SYCOPHANTA : aureo-nitens, thorace cœruleo, elytris aureo-viridibus ftriatis, thorace fubatro. *Linn. Fn. Suec.* 790.—*Gmel. Linn. Syft. Nat.* 1966. *n.* 12.—*Geoffr. Inf. p.* 1. *p.* 144. *n.* 5.—*Reaum. Inf.* 2. *t.* 37. *f.* 18.—*Sulz. Hift. Inf. t.* 7. *f.* 1. —*Bergftr. Nomencl.* 1. t. 12. *f.* 1. 2.

———

One of the largeft and moft fplendid of the European Carabi, and which has not, till very lately, been difcovered in England. It is mentioned, in the firft inftance, by Dr. Turton, as a Britifh fpecies,

E 2   and

and has, fince that time, been met with by entomological collectors, both in Norfolk and Ireland.　Mr. Hooker, F. L. S. poffeffes an example taken in England.

FIG II. II.

CARABUS CRUX MAJOR.

LARGER CRUCIATE CARABUS.

SPECIFIC CHARACTER

AND

SYNONYMS.

Thorax and head black and downy : wing-cafes ferruginous, with a black crofs.

CARABUS CRUX-MAJOR : thorace capitque nigro-villofa, coleoptris ferrugineis : cruce nigra.　*Linn. Syft. Nat.* 673. 39.—*Faun. Suec.* 808.—*Fabr. Ent. Syft.* 1. a. 160. 158.—*Gmel.* 1978. 39.—*Marfh. Ent. Syft. T.* 1. *p.* 471.

CARABUS BIPUSTULATUS.　*Fabr. Syft. Ent.* 207. 59.—*Sp. Inf.* 1. 312. 74.

Le Chevalier noir.　*Geoffr.* 1. 150. 17.

Bupreftis cruciata.　*Panz. Foet.* 2. 70. 7. *t.* 34. *f.* 7.

An infect of elegant formation, very beautiful in colour, and of the greater intereft to the Englifh naturalift, as being rare.　The black cruciate mark on the red wing-cafes conftitute a character of much fingularity.

PLATE CCCCLXXVII.   29

fingularity. The fpecies is of a moderate fize, or rather fmall, and appears to peculiar advantage when magnified.

The fmaller figure on the blade of grafs in the upper part of the plate exhibits the natural fize; the magnified figure is enlarged to about the magnitude of Carabus Sycophanta, a fize which admits of its being depicted with the greater fidelity.

PLATE

478

PLATE CCCCLXXVIII.

ICHNEUMON BILINEATOR.

BILINEATED ICHNEUMON.

HYMENOPTERA.

GENERIC CHARACTER.

Mouth with a ſtraight horny membranaceous bifid jaw, the tip rounded and ciliated : mandibles curved, ſharp ; lip cylindrical, membranaceous at the tip, and emarginate : feelers four, unequal, filiform, and ſeated in the middle of the lip : antennæ ſetaceous, of more than thirty articulations : ſting exſerted, incloſed in a cylindrical ſheath compoſed of two valves, and not pungent.

SPECIFIC CHARACTER.

ICHNEUMON BILINEATOR : black : two incurvate yellow lines on the
head : ſcutel and antennæ in the middle whitiſh.
ICHNEUMON MOLITORIUS *var.* ?

This curious infeċt reſembles, in a very peculiar degree, the Ichneumon molitorius, from which it is, however, diſtinguiſhed by its ſuperiority in ſize, and the two yellowiſh lines on the back part of the head : theſe lines are placed between the eyes as remotely as poſſible, each forming a marginal fillet, which partially ſurrounds the contiguous eye. We ſcarcely feel authorized in the perſuaſion of its being only a variety of the above-mentioned infeċt, although, from its general

aſpeċt,

aſpect, this opinion does not appear altogether improbable : to us it ſeems rather a diſtinct ſpecies than variety. Many examples of Ichneumon molitorius have occured to our own obſervation, but we have never perceived in any of theſe the ſlighteſt trace of the yellow lines, ſo conſpicuous on the head of the preſent inſect.

The ſpecimen, from whence the above figure is taken, is the only one of its kind with which we are acquainted.

P L A T E

479

PLATE CCCCLXXIX.

FIG. I.

PHALÆNA PECTINATARIA.

GREEN CARPET MOTH.

LEPIDOPTERA.

GENERIC CHARACTER.

Antennæ gradually tapering from the bafe ; wings in general deflected when at reft. Fly by night.

SPECIFIC CHARACTER.

PHALÆNA PECTINATARIA. Anterior wings greenifh, with bafe, and two denticulated bands darker : two fufcous V-like marks on the coftal margin, and fufcous fpot near the tip : pofterior wings with a band of dots below the middle.

PHALÆNA PECTINATARIA. *Marfh. M.S.*

One of the moft frequent of the moth tribe, diftinguifhed by the name of " Carpets."

VOL. XIV. F FIG.

FIG. II.

PHALÆNA RUPTATA.

BROKEN BAR, or HORNSEY CARPET MOTH.

SPECIFIC CHARACTER.

PHALÆNA RUPTATA. Anterior wings fubteftaceous : bafe, interrupted broad band in the middle, and fpot at the
tip fufcous, jagged, and margined with white :
pofterior wings pale, with central dot.
GEOMETRA RUPTATA. *Hüb. Schmet. Geom.* 57. 295.—*Sepp.
p.* 11. *pl.* 14 ?

An elegant and by no means abundant fpecies, found in the woods
during the month of June. This infect appears to be rather local,
and from being ufually taken by collectors in the woods of Hornfey,
has long fince obtained among them the trivial appellation of **the
Hornfey Carpet Moth.**

FIG

PLATE CCCCLXXIX. 35

FIG. III.

PHALÆNA MIATA.

AUTUMN GREEN CARPET.

SPECIFIC CHARACTER.

PHALÆNA MIATA. Wings grey-green, with three greenifh bands; the middle one waved with brown: pofterior wings pale, with faint fcalloped bands, and central dot.

PHALÆNA MIATA: alis grifeis: fafciis tribus viridibus: inter media latiore fufco undata. *Linn. Syfl. Nat.* 2. 869. 249.—*Clerk. Icon. pl.* 8. *fig.* 2.

PHALÆNA MIATA. *Fab. Ent. Syfl.* 3. 180. 183.

———

Appears in the winged ftate late in Autumn, whence it has obtained the name of Autumn Green Carpet. The fpecies varies in point of colouring as well as fize, and alfo feems to be very local, if not rare. Among the collectors near London, it is rather better known by the title of Dartford Green Carpet, (from being met with chiefly in the woods adjacent to the town of Dartford, in Kent) than by that of Autumn Green Carpet.

F 2 PLATE

PLATE CCCCLXXX.

CARABUS MELANOCEPHALUS.

BLACK HEADED CARABUS,

GENERIC CHARACTER.

Antennæ filiform : feelers generally fix, the laft joint obtufe and truncated : thorax flat, and margined : wing-cafes marginate.

SPECIFIC CHARACTER

AND

SYNONYMS.

Thorax and legs ferruginous : head and wing-cafes black.

CARABUS MELANOCEPHALUS: thorace pedibufque ferrugineis, elytris capiteque atris. *Linn. Fn. Suec.* 795.— *Gmel. Linn. Syft. Nat.* 1973. *n.* 22.—*Fabr. Sp. Inf.* 1. *p.* 310. *n.* 64.—*Mant. Inf.* 1. *p.* 202. *n.* 89.—*Marfh. Ent. Brit.* 1. 438. 15.
Bupreftis dorfo rubro. *Panz. Voet.* 2. 73. 15.
Le Buprefte noir à corcelet rouge. *Geoff.* 1. 162. 42.

The fmall figure in the annexed plate denotes the natural fize. Linnæus defcribes it as a fylvan fpecies. We met with it in plenty in the woods of Erdig, Denbighfhire.

PLATE

PLATE CCCCLXXXI.

PAPILIO ARGIOLUS.

AZURE BLUE BUTTERFLY.

LEPIDOPTERA.

GENERIC CHARACTER.

Antennæ clubbed at the end: wings erect when at reſt. Fly by day.

SPECIFIC CHARACTER

AND

SYNONYMS.

Wings without a tail: above blue, with black margin: beneath blueiſh, with black dots.

PAPILIO ARGIOLUS: alis ecaudatis ſupra cœruleis margine nigris, ſubtus cœrulefcentibus: punctis nigris difperfis. *Linn. Fn. Suec.* 1076.—*Gmel. Syſt. Nat. T.* 1. *p.* 5. 2350. 234.
HESPERIA ARGIOLUS. *Fabr. Spec. Inf.* 2. *p.* 123. *n.* 551.— *Mant. Inf.* 2. *p.* 73. *n.* 686.

Papilio Argiolus is a very beautiful ſpecies: the female, which is rather larger than the male, is of a vivid azure blue on the upper ſurface; the female blueiſh, inclining to purple: the under ſurface in both are very nearly ſimilar.

The

The larva of this butterfly is rarely met with : in the fly ftate the fpecies is not uncommon, appearing about the middle of the day, in funny weather, on the fkirts of meadows : one brood in the month of June or July, and another the latter end of Auguft. The larva is to feed on grafs.

PLATE

482

PLATE CCCCLXXXII.

GRYLLUS RUFUS.

RUFOUS GRASSHOPPER.

HEMIPTERA.

GENERIC CHARACTER,

Head inflected, armed with jaws: feelers filiform: antennæ setaceous or filiform: wings four, deflected, convolute: the lower ones plaited: hind legs formed for leaping: claws double on all the feet.

SPECIFIC CHARACTER

AND

SYNONYMS.

Thorax cruciate: body fuscous: abdomen rufous: antennæ subclavated and pointed.

GRYLLUS RUFUS: thorace cruciato, corpore rufo, elytris griseis, antennis subclavatis acutis. *Gmel. Linn. Syst. Nat.* 2081. *n.* 56.—Gryllus antennis subclavatis acutis. *Linn. Fn. Suec* 629.

GRYLLUS FUSCUS: abdomine rufo, antennis subclavatis. *Fabr. Sp. Inf.* 1. *p.* 371. *n.* 48.—*Mant. Inf.* 1. *p.* 239. *n.* 55.

Schæff. Icon. tab. 136. *fig.* 4, 5 ?

The ftructure of the antennæ in this fpecies of grafshopper is altogether fingular and characteriftic, the extreme end being dilated into
a pretty confiderable capitulum of a compreffed fubovate form, terminating in nearly an acute point; and which at the firft view bear
a ftrong refemblance to the antennæ of certain fpecies of the papiliones. Appearances of this kind are rare in the Gryllus tribe; the
fpecies clavicornis, a native of Surinam, has antennæ nearly correfponding, and we poffefs another Britilh fpecies, the antennæ of which
are conftructed in a fimilar manner.

We muft acknowledge, that it appears fomewhat anomalous to place
thefe infects with clavated antennæ, among the true Grylli, one decifive character of which confifts in the antennæ being filiform; notwithftanding their fimilitude in other particulars, they might, perhaps,
with far more propriety, conftitute a diftinct genus.

This fpecies is of the fize reprefented, and it is to be obferved, that
the antennæ in one fex is larger than in the other.

Gryllus rufus is defcribed as being very common in fterile fields
in various parts of Europe : on the banks in the Batterfea meadows,
near the river, it is obferved in fome abundance during the month of
September, as we are informed by Mr. Leach.

PLATE

483

PLATE CCCCLXXXIII.

SIREX DROMEDARIUS.

DROMEDARY SAW-FLY.

HYMENOPTERA.

GENERIC CHARACTER.

Mouth with a thick horny truncated fhort denticulated mandible : feelers four, the pofterior ones longer and thicker upwards : antennæ filiform, of more than twenty-four equal articulations : fting exferted, ferrated, ftiff : abdomen feffile, terminating in a point : wings lanceolate, incumbent, the lower ones fhorter.

SPECIFIC CHARACTER

AND

SYNONYMS.

Abdomen black, rufous in the middle, with a white dot on the fide of each fegment : fhanks white at the bafe.

SIREX DROMEDARIUS : abdomine atro : medio rufo ; puncto untrinque albo, tibiis bafi albis. *Fabr. Ent. Syft. T. 2. p.* 128. 16.—*Rofs. Fn. Etr.* 2. 34. 737.— *Gmel.* 2673. 5.

● 2   This

This elegant little infect is moft accurately and minutely defcribed by Fabricius*, from a fpecimen taken at Kiel, in Pruffia, and preferved in the cabinet of Daldorf. According to Roffius, it is alfo a native of Italy. We believe the fpecies has not been before noticed as an inhabitant of Britain.

Our drawings are taken from a fpecimen in the cabinet of Mr. W. Leach, F. L. S. The fmaller figure denotes the natural fize.

It fhould be obferved, that the antennæ do not ftrictly agree with thofe of the Sirex genus in general, the joints being fewer in number, and exhibiting alfo fome lefs material difference in their general ftructure.

---

* Statura & fumma affinitas S. Cameli. Caput globofum, nigrum lineolis duabus verticalibus albis. Thorax antice anguftatus, niger puncto ante alas albo. Alæ obfcuræ. Abdominis fegmentum 1, 2 nigra, 3, 4, 5, 6, 7 rufa, 8 nigrum macula utrinque alba, 9 nigrum, immaculatum.

P L A T E

PLATE CCCCLXXXIV.

CARABUS CEPHALOTES.

COLEOPTERA.

GENERIC CHARACTER.

Antennæ filiform : feelers generally fix, the laft joint obtufe and truncated : thorax flat and margined : wing-cafes marginate.

SPECIFIC CHARACTER

AND

SYNONYMS.

Deep black, thorax attenuated behind, the pofterior margin rugofe with dots : wing-cafes fmooth, and fcarcely ftriated.

CARABUS CEPHALOTES : ater, thorace poftice attenuato, margine poftico punctato-rugofo, elytris lævibus obfoletiffimè ftriatis. *Marfh. Ent. Brit. T.* 1. 472. *n.* 107.

CARABUS CEPHALOTES : apterus, elytris atris lævibus, thorace exferto oblongo. *Linn. Fn. Suec.* 788.—*Gmel. Linn. Syft. Nat. T.* 1. *p.* 4. 1964. 9.

CARABUS CEPHALOTES : apterus ater læviffimus, thorace orbiculato convexo. *Fabr. Sp. Inf.* 1. *p.* 304. *n.* 27. —*Mant. Inf.* 1. *p.* 198. *n.* 39.

SCARITES CEPHALOTES. *Panz. Ent. Germ.* 37. 5.

Pfeudocupis major. *Panz. Voet.* 2. 64. 2. *t.* 33. *f.* 2.

Found on fandy fhores of the fea.

PLATE

PLATE CCCCLXXXV.

FIG. I. II.

PHALÆNA LINEATARIA.

PALE TRIPLE-BAR MOTH.

GENERIC CHARACTER.

Antennæ gradually tapering from the bafe to the tip : tongue fpiral : wings in general deflected when at reft.   Fly at night.

* GEOMETRA.

SPECIFIC CHARACTER.

PHALÆNA LINEATARIA.   Pale : anterior wings with an oblique bilineated band at the bafe : trilineated band near the tip : bar in the middle angulated, and inclofing a dot near the coftal margin : pofterior pair fubli-neated : exterior margin of all the wings dotted.

———

This we are inclined to confider as an extremely rare fpecies.   The fpecimen reprefented in the upper part of the plate, and to which the figure I. is annexed, will be obferved, at the firft view, to differ from that fhewn beneath at figure II. in the diftinctnefs of its markings; but this alone feems to conftitute their real difference, as every lineation in the lower fpecimen accords with thofe exhibited in the infect fhewn above.   The latter appears to be either a pale variety, or an example of the fpecies in lefs perfect condition than the other.   Both infects are fhewn in their natural fize.

FIG.

FIG. III.

PHALÆNA RUBRO-VIRIDATA.

BULLSTRODE GREEN CARPET MOTH.

SPECIFIC CHARACTER.

PHALÆNA RUBRO-VIRIDATA.   Anterior wings greenish, tinged
    with rufous : bafe and broad band in the middle
    fubfufcous : pofterior wings brownifh.
PHALÆNA RUBRO-VIRIDATA.   *Marfh. M. S.*
PHALÆNA PSITTACATA.   *Fabr. Ent. Syft.* 3. 195. 238 ?

———————

Occurs in the winged ftate in the month of October.

We are not without fufpicion, that the moth reprefented in that
fcarce work, the " *Icones*" of Clerk, (fig. 8. pl. 4.) may be intended
for an infect of this fpecies.   The figure appears without any
name.

P L A T E

PLATE CCCCLXXXVI.

CARABUS CREPITANS.

MUSKETEER BEETLE.

GENERIC CHARACTER.

Antennæ filiform : feelers generally fix, the laft joint obtufe, and truncated : thorax flat and margined : wing-cafes marginate.

SPECIFIC CHARACTER

AND

SYNONYMS.

Head, thorax, and legs ferruginous : wing-cafes blue-black.

CARABUS CREPITANS: capite thorace pedibufque ferrugineis, elytris
nigris. *Linn. Syft. Nat.* 671. 18.—*Fn. Suec.*
792.—*Fabr. Syft. Ent.* 242. 35. *Sp.* 1. 307.
44.—*Mant.* 1. 200. 61.—*Panz. Ent. Germ.*
51. 35.—*Oliv.* 3. 35. 64. 80.—*Marfh. Ent.
Brit.* 1. 468. 96.
Le Buprefte à tête, corcelet, et pattes rouges et étius bleus. *Geoffr.*
1. 151. 19.

An infect of fmall fize that inhabits Europe, and is fometimes found in England, where it is far from common.

VOL. XIV.   H   This

P L A T E CCCCLXXXVI.

This fpecies is remarkable only for the peculiar mode of defence which it inftinctively adopts when clofely purfued by carnivorous infects, or other enemies : on thefe occafions, it emits a diftinct, and rather loud noife, either from the vent, or, as fome fuppofe, from the friction of the wing-cafes. This found it has the ability to repeat feveral times, and which, it may be imagined, is feldom exerted without fuccefs; the unexpected explofion for the moment alarming or repulfing its purfuer, and allowing, by that means, a convenient interval for the infect purfued to effect its efcape.

An enlarged figure of this infect is given with its natural fize,

PLATE

487

51

PLATE CCCCLXXXVII.

FIG. I. I.

PHALÆNA TESTACEATA.

PALE SCALLOP MOTH.

LEPIDOPTERA.

GENERIC CHARACTER.

Antennæ gradually tapering from the bafe to the tip : tongue fpiral; wings in general deflected when at reft. Fly at night.

* GEOMETRA.

SPECIFIC CHARACTER.

PHALÆNA TESTACEATA. Whitifh, with numerous teftaceous fcalloped lines : a common broad pale band in the middle ; and marginal feries of oblong black dots.

———

The infect from whence the above defcription and annexed figures are taken, is the only example of its fpecies we have feen, and hence we are inclined to confider it exceedingly fcarce, if not perfectly unique. The fmaller figure exemplifies the natural fize.

N 2

FIG.

FIG. II.

PHALÆNA CUNEATA.

CUNEATE MOTH.

SPECIFIC CHARACTER.

PHALÆNA CUNEATA. Anterior wings fufcous, with two pale
broad bands, the inner one angulated, and the
exterior marked in the middle with a fingle
feries of cuneate fufcous fpots.

―――――

A fpecies of very ftriking appearance, and fufficiently diftinguifhed
by the feries of wedge-formed fpots difpofed along the pale exterior
band of the upper wings. The fufcous ground colour forms a pretty
broad and diftinct band in the middle of the wings, and is further cha-
racterized by an oblong, and fomewhat paler fpot, contiguous to the
anterior margin, as well as a geminous or rather bipupillate fpot at the
pofterior edge of the fame band. The lower wings are whitifh, with
pale fufcous fcalloped marginal lines, and a dufky dot in the middle.

P L A T E

488

PLATE CCCCLXXXVIII.

CARABUS COMPLANATUS.

SAND CARABUS.

COLEOPTERA.

GENERIC CHARACTER.

Antennæ filiform: feelers generally fix, the laft joint obtufe and truncated: thorax flat and margined: wing-cafes marginate.

SPECIFIC CHARACTER

AND

SYNONYMS.

Pale: two black-waved lines on the wing-cafes.

CARABUS COMPLANATUS: pallidus, elytris fafciis duabus undulatis nigris. *Linn. Syft. Nat.* 2. 671. 17.

CARABUS ARENARIUS: pallidus elytris maculis duabus dorfalibus atris. *Fabr. Spec. Inf.* 1. 305. 34.—*Syft. Ent.* 241. 26.—*Mant. Inf.* 1. *p.* 199. *n.* 46.

———

The very elegant and interefting fpecies of Carabus, at prefent before us, appears to be the original C. Complanatus of Linnæus: this we learn from the authentic fpecimen of that infeÆ defcribed by Linnæus himfelf, and which, conftituting a part of the Linnæan cabinet, is now in the poffeffion of Dr. Smith.

The

The fame infect is, beyond difpute, the genuine Carabus arenarius of Fabricius, as may be clearly afcertained from the original example of that fpecies defcribed by Fabricius in the Bankfian cabinet.   Fabricius was doubtlefs not aware that it had been previoufly defcribed, and therefore, from its habits of life, very appofitely affigned it the fpecific name of arenarius : the Linnæan name, however, deferves the preference in point of priority, and, being perfectly admiffible, fhould in candour be retained.

Linnæus, perhaps on authority not fufficiently explicit, fpeaks of his fpecies Complanatus as an inhabitant of the ifland of St. Domingo. It is poffible, his information in this refpect might be correct, but we are rather inclined to think it doubtful. Its exiftence, as a Britifh fpecies, is determined in the moft conclufive manner.

It will not be improper to obferve, that the firft example of this fpecies, difcovered in Britain, was taken, fome years ago, by Sir Jofeph Banks on the fandy fhores of Wales, a circumftance to which Fabricius adverts, though flightly.   From the time of its difcovery, we have reafon to believe it was not again obferved till within a very recent period, when, on further fearch about the fame fhores where it was firft obferved, it was again found, and in confiderable plenty.   During the fummer of the year 1809, it was taken in abundance under the driftwood on the fhores near Cromllyn Burrows, in the vicinity of Swanfea, by Mr. W. Leach, F. L. S. ; and prior to that period, Mr. L. W. Dillwyn, F. L. S. met with it on the fands below the town of Newton, in Glamorganfhire.

In a living ftate, this curious infect appears uncommonly pellucid, and this appearance is retained in a certain degree even in the examples dried, and prepared for the cabinet: the general colour is pale teftaceous, or yellowifh, with the extreme tips of the jaws and eyes brown.   The two black or deep brown fpots on the back conftitute diftinct denticulated bands ; and two or more of the longitudinal ftriæ, which interfect the pale tranfverfe band between thofe fpots, are

likewife

likewife black. The whole of the lower furface, with the legs and antennæ, are pale yellowifh teftaceous.

In conclusion we ought to mention, that this infect varies materially in the form, and alfo in the intenfity of the black or dufky marks on the wing-cafes.

P L A T E

PLATE CCCCLXXXIX.

APIS MANICATA.

MANICATED BEE.

HYMENOPTERA.

GENERIC CHARACTER.

Mouth horny: jaw and lip membranaceous at the tip: tongue in-flected: feelers four, unequal, and filiform: antennæ fhort and filiform in the male, in the female fubclavated: wings flat: fting of the fe-males and neuters pungent, and concealed in the abdomen.

SPECIFIC CHARACTER

AND

SYNONYMS.

Cinereous, abdomen black, with yellow lateral fpots: tail armed with five teeth.

APIS MANICATA: cinerea, abdomine nigro, maculis flavis lateribus, ano quinque dentato. *Fabr. Ent. Syjt. n.* 73.

APIS MANICATA: nigra, pedibus anticis hirfutiffimis, abdomine ma-culis lateribus, ano tridentato. *Linn. Syjt. Nat.* 12. *n.* 28.—*Fn. Suec.* 1701. *Fourcroy. Ent. Par. n.* 3. *Geoff. Hijt. Inf. Par.* 2. 408. *n.* 3. *Kirby. Ap. Angl. V.* 2. 248. 47.

VOL. XIV.   I   The

The five diftinct denticles at the extremity of the abdomen form an excellent fpecifical diftinction of this kind of bee. The fpecies is very common in fome parts of Britain. When on the wing, it is obferved to hover over flowers in the fame manner as Sphinx Stellatarum : the Glechoma hederacea (ground ivy) appears to be its favourite, being found during the greater part of the fummer on beds of thefe fragrant plants.

When the female prepares to conftruct the nidus in which the infant brood is to be depofited, fhe feeks a convenient hollow in old palings, the cavity of a wall, or other retreat eligible for her reception ; and having determined the fpot, fhe next reforts to fome tomentous or woolly kind of plant, to obtain materials for the completion of her object. The portion of down required fhe ftrips or fhaves off with aftonifhing celerity and addrefs, conveys it away to her hiding-place in bundles between her head and fore legs, and repeats her vifits till the quantity procured prove fufficient for her ufe. She then proceeds to line the infide of the cavity with the down, and lays her eggs, each of which is enveloped in a feparate covering, compofed of the fame vegetable materials.

Some accurate obfervers of the habits of this induftrious little infect have been led to imagine, that it employs only the tomentum or down of one particular kind of plant, namely, that of Agroftemma coronaria ; and it does indeed appear, from the refult of their remarks, that the nidus is in general conftructed with the down of this fpecies of vegetables. There is neverthelefs fome reafon for believing, that the down collected for this purpofe is not on every occafion confined exclufively to the plant before mentioned.

P L A T E

PLATE CCCCXC.

FIG. I.

MUSCA INANIS.

DIPTERA.

GENERIC CHARACTER.

Mouth with a foft exferted flefhy probofcis, and two equal lips : fucker furnifhed with briftles : feelers two, and fhort, or fometimes none : antennæ generally fhort.

SPECIFIC CHARACTER

AND

SYNONYMS.

Brown : abdomen pale yellow, with three black bands.

MUSCA INANIS : antennis plumatis pilofa flavefcens, abdomine pel-
        lucido cingulis duobus nigris. *Linn. Syft. Nat.
        XII. 2. p. 989. n. 61.—Fn. Suec. 1825.*
SYRPHUS INANIS : fufca, abdomine pellucido : cingulis tribus nigris.
        *Fabr. Spec. Inf. 1. p. 435. n. 1.—Mant. Inf.
        1. p. 342. n. 1.*
Mufca apivora. *Degeer. Inf 6. p. 56. n. 3. t. 3. f. 4.*
Volucella fexta. *Schæff. Icon. pl. 36. fig. 7. 8.*

This is an interefting fpecies, and not common : the figure denotes the natural fize.

I 2

FIG.

FIG. II.

MUSCA HIRSUTA.

SPECIFIC CHARACTER.

Deep black, glofſy, and befet with long briftly hairs : wings blackiſh, at the bafe ſubfuſcous.

Musca tremula. *Fabr. Spec. Inſ. 2. p. 442. n. 32?*

━━━━━━━

The prefent infect bears a ftrong refemblance to the Mufca groſſa, of which it might be confidered, at the firft view, as a dwarf variety, being rather leſs than half the fize of that fpecies. As in Mufca groſſa, the thorax and abdomen are befet with ftiff briftly hairs, but thefe are more numerous, and at leaft twice the length in proportion, in the prefent fpecies, to thofe on the former infect.

The Mufca hyftrix of Drury is very fimilar to this in appearance, but is larger : it approaches, however, ftill nearer the infect called by Harris (Expof.) Mufca obfidianus, than Mufca hyftrix.

From the cabinet of Dr. Letfom.

P L A T E

491

PLATE CCCCXCI.

CERAMBYX CORIARIUS.

LARGE ELM CERAMBYX.

COLEOPTERA.

GENERIC CHARACTER.

Antennæ fetaceous: eyes lunate, and embracing the bafe of the antennæ: feelers four: thorax fpinous, or gibbous: wing-cafes lınear: body oblong.

SPECIFIC CHARACTER

AND

SYNONYMS.

Thorax three-toothed: body pitchy: wing-eafes mucronate: antennæ fhorter than the body.

CERAMBYX CORIARIUS: thorace tridentato, corpore piceo, elytris mucronatis, antennis brevioribus. *Linn. Syft. Nat.* 622. 7.—*Fn. Suec.* 647.—*Gmel.* 1815. 7. *Marfh. Ent. Brit. T.* 1. *p.* 325. 1.

PRIONUS CORIARIUS. *Fabr. Syft. Ent. t.* 24. *f.* 4.—*Spec. Inf.* 1. 206. 9 —*Mant.* 1. 129. 13.—*Ent. Syft. i. b. Panz. Faun. Germ.* 9. *t.* 8.

Cerambyx Prionus. *Degeer, v.* 59 1. *t.* 3, *f.* 5.

Le Prione, *Geoff.* 198. 1. *t.* 3. *f.* 5.

Both

Both fexes of this curious beetle are reprefented in the annexed plate, the male in the attitude of crawling on the ground, the female in the act of flight. The female is rather larger than the male, and has the antennæ of a more fetaceous form. The antennæ of the other fex are remarkable for their magnitude, and contribute very materially to the interefting appearance of the infect.

Cerambyx Coriarius is the moft confpicuous infect, in point of fize, among the Britifh cerambyces, and is always confidered as a fcarce and valuable fpecies. It is found chiefly in decayed wood, more efpecially in the trunks of rotten elms.

P L A T E

492

PLATE CCCCXCII.

APIS MELLIFICA.

COMMON BEE.

HYMENOPTERA.

GENERIC CHARACTER.

Mouth horny : jaw and lip membranaceous at the tip : tongue in-flected : feelers unequal, and filiform : antennæ fhort and filiform in the males : in the female fubclavated : wings flat : fting of the females and neuters pungent, and concealed in the abdomen.

SPECIFIC CHARACTER

AND

SYNONYMS.

Pubefcent : thorax greyifh : abdomen brown : pofterior fhanks ciliated and tranfverfely ftriate within.

APIS MELLIFICA : pubefcens, thorace fubgrifeo, abdomine fufco, tibiis pofterioribus ciliatis : intus tranfverfe ftriatis. *Linn. Fn. Suec.* 1697.—*Fabr. Sp. Inf.* 1. 2. 480. *n.* 37.—*Mant. Inf.* 1. *p.* 302. *n.* 42.
Apis domeftica five vulgaris. *Ray. Infect. p.* 240.
Apis gregaria. *Geoff. Inf. Par.* 2. *p.* 407. *n.* 1.
Reaum. Inf. 5. *Tab.* 21, 22, 23.

The

The Common Honey Bee is rarely found in a wild ftate in Britain: fuch as occur in this ftate of nature build netts in the hollows of de- cayed trees, which they inhabit in large focieties, and are faid to ob- ferve the fame order and policy in the regulation of their community as when domefticated in the hive. The figures in the annexed plate are from examples difcovered wild.

The two upper figures reprefent the male and female, that in the lower part of the plate is the figure of the neuter. The male or drone is diftinguifhed by having the eyes remarkably large, and approximate behind, and alfo by the abdomen being robuft, and fomewhat obtufe ; in the female, or queen bee, the eyes are fmall and remote, the wings fmaller, and the abdomen remarkably large, elongated, and conic. The neuters are the working bees, and it is the office of thofe induf- trious creatures to collect the nectareous juices of flowers for making honey and wax, to feed and protect the young, and defend their fociety againft every affailant.

PLATE

493

PLATE CCCCXCIII.

FIG. I.

PHALÆNA BERBERATA.

BARBERRY MOTH.

LEPIDOTERA.

GENERIC CHARACTER.

Antennæ gradually tapering from the bafe: wings in general deflected when at reft. Fly by night.

* GEOMETRA.

SPECIFIC CHARACTER

AND

SYNONYMS.

Anterior wings cinereous and fubrufous, with three brown lineate bands, the pofterior one inclofing a pale ε.

PHALÆNA BERBERATA: feticornis alis anticis cinereis: fafciis tribus fafcis: pofteriori nigro undata. *Fabr. Mant. Inf. T. 2. p.* 203. *n.* 154.—*Ent. Syft. T. 3. p. 2.* 182. 189.

GEOMETRA BERBERATA. Der Gauerdorn Spanner. *Wiener Verz. p.* 113. *et No.* 23.

GEOMETRA BERBERATA Jungs Alphabet. *Berzeichn.* 1. *p.* 75. —11. *p.* 370.

PHALÆNA BERBERATA. Der Gauerdorn Spanner. *Kleem.*
Beytr. Naturf. Inf. Gefch. p. 32. n. 9.

———

This pretty Moth is produced, according to Fabricius, from a fca-
brous larva of a brown colour, varied with rufous and white, and
which, according to the continental writers in general, as well as Fabricius, is found on the common barberry, *berberis vulgaris*. The
larva we have not feen, but, from a minute defcription and plate in the
latter part of the Supplement of Kleeman's *Beytraege*, we are enabled
to fpeak of it in more precife terms than Fabricius, and alfo to defcribe
its pupa. The larva is of the looper kind, and rather thick in proportion to the length; of a brownifh colour, with black dots, and
two fhort black parallel lines at the pofterior extremity, extending the
length of the three or four laft fegments. The pupa is chefnut brown,
rather inclining to an ovate form, and is inclofed in a fpinning woven
between two or three leaves, which are drawn nearly together for that
purpofe.

The fpecies occurs in the winged ftate, as a native of Britain, in
feveral cabinets, though we have never underftood it to be common.
We have named it the Barberry Moth, in allufion to the plant on
which the larva ufually feeds : among the Englifh collectors, it bears
two or more indefinite appellations.

FIG.

FIG II.

PHALÆNA RUMIGERATA.

SCALLOPED-WING FOUR-DOT MOTH.

SPECIFIC CHARACTER.

PHALÆNA RUMIGERATA. Wings deeply angulated, produced behind, and fcalloped : pale teftaceous, with two
tranfverfe dark lines on the anterior wings, and
one on the pofterior : a fufcous dot in the middle
of each wing.

———————

The elongated form of the wings, and depth of the remarkably produced pofterior extremity of the lower pair, fufficiently diftinguifh this
from the following fpecies. The example, from which the above
figure is taken, appertains to the colleƈtion of the late Mr. Drury.

K 2   FIG.

FIG. III.

PHALÆNA QUADRIPUNCTATA.

QUADRIPUNCTATE MOTH.

SPECIFIC CHARACTER.

PHALÆNA QUADRIPUNCTATA. Wings fubangulated: fomewhat teftaceous, with a common line near the bafe, dot in the middle, and common line behind.

From the fame cabinet as the preceding.

PLATE

PLATE CCCCXCIV.

MUSCA HOTTENTOTTA.

DIPTERA.

GENERIC CHARACTER.

Mouth with a soft exserted fleshy proboscis, and two equal lips : suckers furnished with bristles : feelers two, very short, or sometimes none : antennæ generally short.

SPECIFIC CHARACTER

AND

SYNONYMS.

Body covered with yellow hairs : wings hyaline, with fuscous rib.

MUSCA HOTTENTOTTA : hirta flavescens, alis hyalinus : costa fusca. *Linn. Fn. Suec.* 1787.—*Gmel. Linn. Syst. T.* 1. *p.* 5. 2831. 13.—*Fabr. Spec. Inf. 2. p.* 415. *n.* 16.

NEMOTELUS HOTTENTOTTUS. *Degeer. Inf.* 6. *p.* 190. *n.* 12. *t.* 11. *f.* 7.—*Schæff. Icon. t.* 76. *f.* 6.

━━━━━━━

A large, curious, and elegant species, and one of considerable rarity in this country : in the north of Europe it appears to be very far from uncommon.

PLATE

PLATE CCCCXCV.

FIG. I. I.

VESPA ANGULATA.

ANGULATE WASP.

HYMENOPTERA.

GENERIC CHARACTER.

Mouth horny, with a compreffed jaw: feelers four, unequal and filiform: antennæ filiform, the firft joint longer and cylindrical: eyes lunar: body glabrous: upper wings folded in each fex: fting pungent, concealed in the abdomen.

SPECIFIC CHARACTER.

VESPA ANGULATA. Head black: thorax black, with yellow ante. rior margin: abdomen yellow, with triangular black fpot at the bafe, and broad black band iu the middle.

———

This is one of the fmalleft fpecies of the wafp genus: the head and thorax black, except the margin in the front of the latter, which is yellow: the body yellow, with a peculiar angulate or triangular black fpot at the bafe, pointing downwards, and a band of the fame in the middle. The antennæ and thighs are black, legs yellow.

This infect does not appear to have been before defcribed. The fmaller figure denotes the natural fize.

FIG.

FIG. II.

VESPA QUADRATA.

QUADRATE WASP.

SPECIFIC CHARACTER.

VESPA QUADRATA. Head and thorax black, the latter with yellow
anterior margin : abdomen with a fquare fpot of
black at the bafe, and broad black band in the
middle.

VESPA QUADRATA. *Panzer. Inf. Germ.*

———————

Exceeds the former fpecies in fize, and differs in having a quadran-
gular inftead of triangular black fpot at the bafe of the abdomen, and
the thorax marked in the middle with dots of yellow. This fpecies is
not uncommon.

It has not been conceived requifite to add an enlarged figure of this
infect.

P L A T E

496

PLATE CCCCXCVI.

DYTISCUS 12-PUSTULATUS.

12-SPOT WATER-BEETLE.

COLEOPTERA.

GENERIC CHARACTER.

Antennæ fetaceous : palpi fix, and filiform : pofterior legs formed for fwimming: fringed on the inner fide, and nearly unarmed with claws.

SPECIFIC CHARACTER

AND

SYNONYMS.

Teftaceous : wing-cafes black, with fix teftaceous fpots on each.

DYTISCUS 12-PUSTULATUS: teftaceus, elytris nigris: maculis fex teftaceis. *Fabr. Ent. Syft.* 1. *a. p.* 197. 50.
Paykul. Faun. Suec. 1. 220. 29.
Oliv. 3. 40. 31. 35. *t.* 5. *f.* 46. *a. b.*
Marfh. Ent. Brit. 1. *p.* 422. 12.

———————

Few of the Dytifci are diftinguifhed for their beauty : their colours in general are either black, or blackifh, varioufly gloffed with blueifh purple, or olive, or of a dull ferruginous ; and it is for this reafon, more efpecially than any other, that the prefent fpecies claims particular attention : it is certainly one of the prettieft infects of its tribe.

VOL. XIV.   L   The

The fize of this fpecies is inconfiderable, which renders it neceffary, in order to convey a correct idea of the fubject, to reprefent it both in its natural fize, and as it appears when magnified. The colour of the antennæ, legs, and thorax, are teftaceous, the latter marked in the middle with a band of black; the wing-cafes are black, with fix teftaceous fpots of an irregular form, difpofed in two longitudinal feries on each. Like the reft of its tribe, Dytifcus 12-Puftulatus is of the aquatic kind.

P L A T E

497

PLATE CCCCXCVII.

FORMICA RUFA.

RUFOUS ANT.

HYMENOPTERA.

GENERIC CHARACTER.

Feelers four, unequal, with cylindrical articulations, placed at the tip of the lip, which is cylindrical, and nearly membranaceous: antennæ filiform : a fmall erect fcale between the thorax and abdomen : females and neuters armed with a concealed fting : males and females furnifhed with wings, neuters winglefs.

SPECIFIC CHARACTER

AND

SYNONYMS.

Black : thorax compreffed, and with the legs ferruginous.

FORMICA RUFA : nigra, thorace compreffo pedibufque ferrugineis.
   Fabr. Sp. Inf. 1. *p.* 489. *n.* 6.—*Mant. Inf.* 1.
   p. 308. *n.* 7.
FORMICA FUSCA ? *Geoff. Inf. p.* 2. *p.* 428.

Except the Formica herculanea, to which the prefent fpecies bears a ftriking refemblance both in appearance and magnitude, this is one of the largeft fpecies of the ant tribe found in Europe. Like the

L 2   former,

former, it inhabits woods, and refides chiefly in hollow trees. The neuters, as in the reft of the genus, are winglefs.

The figure in the upper part of the plate is magnified, the lower reprefents it in its natural fize.

PLATE

498

PLATE CCCCXCVIII.

PAPILIO ÆGERIA.

SPECKLED WOOD BUTTERFLY.

LEPIDOPTERA.

GENERIC CHARACTER.

Antennæ clavated at the tip : wings erect when at reft. Fly by day.

SPECIFIC CHARACTER

AND

SYNONYMS.

Wings indented, fufcous with yellow fpots : anterior pair with an ocellar fpot on each fide : pofterior ones ocellated above, beneath marked with four dots.

PAPILIO ÆGERIA : alis dentatis fufcis luteo maculatis : utrinque anterioribus ocello, pofterioribus fupra ocellis, fubtus punctis quatuor. *Fabr. Spec. Inf. 2. p. 73. n. 325.—Mant. Inf. 2. p. 37. n. 381. Gmel. Linn. Syft. Nat. T. 1. p. 5. n. 2295. 143. Ray Inf. p. 128. n. 5. Schæff. Icon. t. 75. f. 1. 2. Hübn. Schmet. pl. 40. 181. 2.*

Very

Very common in the lanes leading through woody fituations during the whole fummer, two or three diftinct broods being produced annually. The larva is green, with a white line, and fpinous tail; the pupa greenifh, and bulky in proportion to its length.

In the larva ftate it feeds on graminiferous plants, and is obferved in this ftage from March till the end of June. The firft brood appears in the fly ftate in the month of April, the lateft in Autumn.

Papilio Ægeria is not only one of the moft abundant and generally diffufed fpecies of Papiliones in this country, but appears to be found in plenty throughout the reft of Europe.

P L A T E

499

PLATE CCCCXCIX.

FIG. I. I:

PHALÆNA TRINOTATA.

TRINOTATED MOTH.

LEPIDOPTERA.

GENERIC CHARACTER.

Antennæ gradually tapering from the bafe to the tip : tongue fpiral : wings generally deflected when at reft. Fly by night.

* GEOMETRA.

SPECIFIC CHARACTER.

PHALÆNA TRINOTATA. Very pale teftaceous : anterior wings with two whitifh lobate fpots in the middle, inclofed between two bands of fcalloped lines : exterior margin with a feries of triangular pale fpots, each containing three black dots.

———

An extremely rare and probably unique infect of the Geometra family of Phalæna. It is a fpecies of elegant and very pleafing afpect, though not in any degree remarkable for the gaiety of its colours.

The chain or feries of pale triangular fpots, which extends along the outer margin of the anterior wings, is altogether characteriftic of

this

this fpecies, each of thofe fpaces containing three diftinct black dots, which alfo are difpofed in the form of a triangle.

Our prefent fpecies is of moderate fize, as is expreffed by the fmaller figure: an enlarged view of the fame is fhewn in the upper part of the plate.

FIG. II.

PHALÆNA LITERATA.

LETTERED MOTH.

SPECIFIC CHARACTER.

PHALÆNA LITERATA. Anterior wings dark fufcous, with waved lines, and a black lineole: band in the middle milky, with a black character refembling T

Larger than the preceding, and perhaps no lefs uncommon.   Both are preferved in the cabinet of Mr. Drury.

A moth very nearly allied to the prefent occurs in the fecond part of the work of Sepp, (plate 5.): the general colour differs in being tinged with greenifh, notwithftanding which it may be a variety of this fpecies.

PLATE

500

PLATE D.

FORFICULA GIGANTEA.

GIGANTIC EARWIG.

COLEOPTERA.

GENERIC CHARACTER.

Antennæ fetaceous : feelers unequal, filiform : wing-cafes half as long as the abdomen : wings folded up under the wing-cafes : tail armed with a forceps.

SPECIFIC CHARACTER

AND

SYNONYMS.

Pale : above variegated with black : tail bidentated : forceps extended, in the male armed with two teeth, in the female fhorter, and ferrated within.

FORFICULA GIGANTEA : pallida fupra nigro variegata ano bidentato : forcipe porrecta unidentata. *Fabr. Ent. Syft.* *T. 2. 1. n. 2.*

———

The largeft of the European fpecies of Forficula, if not of the whole genus, and which has not, till very recently, been difcovered in this country : it is taken in fome plenty at Chrift Church, in Hampfhire.

From the figures in the annexed plate, it will be obferved, that the female is confiderably larger than the other fex, and that the difference in the ftructure of their forceps is ftrikingly obvious. The female has been confidered by fome collectors as the Erythrocephala of Fabricius.

PLATE

PLATE DI.

FIG. I. I.

DYTISCUS HERMANNI.

HERMANN'S WATER-BEETLE.

COLEOPTERA.

GENERIC CHARACTER.

Antennæ fetaceous: palpi fix, and filiform: pofterior legs formed for fwimming, fringed on the inner fide, and nearly unarmed with claws.

SPECIFIC CHARACTER

AND

SYNONYMS.

Gibbous: head and thorax ferruginous: wing-cafes truncated, at the bafe ferruginous.

DYTISCUS HERMANNI: gibbus, capite thorace elytrorumque bafi
ferrugineis, elytris truncatis. *Fabr. Syft. Ent.*
232. 14.—*Sp. Inf.* 1. *p.* 295. 19.—*Mant.* 1.
191. 24.—*Gmel. Linn. Syft. Nat.* 1949. 41.—
Marfh. Ent. Brit. T. 1. *p.* 418. 13.

━━━━━━━

Dytifcus Hermanni does not appear to be a very common infect in this country: it is of an amphibious nature, and refides chiefly in the water, like the other fpecies of its tribe.

The fmalleft figure reprefents the natural fize.

M 2

FIG.

FIG. II. II.

DYTISCUS MACULATUS.

SPOTTED WATER-BEETLE.

SPECIFIC CHARACTER

AND

SYNONYMS.

DYTISCUS MACULATUS. Teſtaceous or pale: thorax duſky, with
a pale band: wing-caſes varied with black ſpots,
and lines.

DYTISCUS MACULATUS. *Fab. Spec. Inſ.* 1. 295. 23?

━━━━━━━

A rare ſpecies, ſhewn both in its natural ſize and magnified. The
varieties of this inſect are very numerous, and diſſimilar.

PLATE

PLATE DII.

VESPA CRABRO.

HORNET.

HYMENOPTERA.

GENERIC CHARACTER.

Mouth horny, with compreffed jaw : feelers four, unequal, filiform : antennæ filiform, the firft joint longer and cylindrical : eyes lunar : body glabrous : upper wings folded in each fex : fting pungent, con-ealed in the abdomen.

SPECIFIC CHARACTER

AND

SYNONYMS.

Thorax black : the anterior part rufous and immaculate : incifures of the abdomen with a double contiguous black dot.

Vespa Crabro : thorace nigro : antice rufo immaculato : abdominis
    incifuris puncto nigro duplici contiguo.   *Linn.*
    Syft. Nat. 2. 948. 3.—*Fn. Succ.* 1670.—*Fabr.*
    Ent. Syft. T. 2. *p.* 255.
    Geoff. Inf. 2. 368. 1.
    Schæff. Icon. 53. *f.* 5. *tab.* 136. *fig.* 3.
    Reaum. Inf. 4. *tab.* 10. *fig.* 9.
    Mouffet. Inf. 50.

The

The Hornet lives in focieties, conftructing its neft in the trunks of hollow trees, or among timber, wood-lofts, and other fimilar fituations: the neft is capacious, being adapted for the reception of a numerous family, though fmaller than that of the common wafp, and containing a lefs confiderable number of cells: its texture refembles that of parchment, or ftrong paper.

In its manners of life the Hornet refembles the wafp, being, like that infect, fierce, voracious, and fubfifting on frefh animal fubftances, on fruits, and the nectar of flowers; and in particular committing vaft havoc among the lower tribes of infects. Its animofity towards the bee is well known: it often enters the hive of thefe induftrious creatures, and plunders them of their honey with impunity.

Both fexes of this fpecies are fhewn in the lower part of the plate, the fpecimens felected for which purpofe rather exceed the ufual magnitude. The figure in the upper part of the plate is the reprefentation of a fmall variety of the fame fpecies.

P L A T E

PLATE DIII.

FORMICA RUBRA.

RED ANT.

HYMENOPTERA.

GENERIC CHARACTER.

Feelers four, unequal, with cylindrical articulations, placed at the tip of the lip, which is cylindrical, and nearly membranaceous: antennæ filiform : a fmall erect fcale between the thorax and abdomen : females and neuters armed with a concealed fting : males and females furnifhed with wings, neuters winglefs.

SPECIFIC CHARACTER

AND

SYNONYMS.

Teftaceous : eyes and dot under the abdomen black.

Formica rubra : teftacea, oculis punctoque fub abdomine nigris. *Linn. Fn. Suec. 2. n.* 1725.—*Fabr. Sp. Inf.* 1. *p.* 490. *n.* 9.—*Mant. Inf.* 1. *p.* 308. *n.* 11.— Formica minima rufa, *Ray.*

Inhabits woods, and lives in nefts under ftones. During the winter they remain torpid, like moft others of this genus.

PLATE

PLATE DIV.

FIG. I.

CARABUS INQUISITOR.

COLEOPTERA.

GENERIC CHARACTER.

Antennæ filiform : feelers generally fix, the laft joint obtufe and truncated : thorax flat, and margined : wing-cafes marginate.

SPECIFIC CHARACTER

AND

SYNONYMS.

Wing-cafes ftriated, braffy-green, with three rows of dots.

CARABUS INQUISITOR : elytris ftriatis viridi-æneis : punctis triplici
ordine. *Linn. Syft. Nat.* 669. 11.—*Fn. Suec.*
789.—*Gmel. Syft. Nat.* 1965. 11.
Fabr. Syft. Ent. 239. 18.—*Spec. Inf.* 1. 303. 23.
Mant. 1. 197. 31.
Paykull. Monogr. 39.—*Fn. Suec.* 1. 127. 40.
Panzer. Ent. Germ. 54. 50.
Marfh. Ent. Brit. T. 1. *p.* 448.
Le Buprefte quarré couleur de bronze antique. *Geoff.* 1. 145. 6.
Bupreftis Sycophanta minor. *Panz. Voet.* 2. 86. 39. *t.* 38. *f.* 39.

─────────

A rare fpecies in Britain : it has been taken in Norfolk.

FIG. II.

CARABUS ROSTRATUS.

SNOUTED CARABUS.

SPECIFIC CHARACTER

AND

SYNONYMS.

Apterous: wing-cafes fmooth and black: thorax narrow: head very flender.

CARABUS ROSTRATUS: apterus, elytris læviufculis nigris thorace anguftiori, capite anguftiffimo. *Fabr. Syft. Ent.* 240. 21.—*Spec. Inf.* 1. 304. 26.—*Mant.* 1. 198. 36.—*Ent. Syft.* 1. *a.* 131. 31.
Paykull. Monog. 26. 14.
Marfh. Ent. Brit. T. 1. *p.* 470.

TENEBRIO ROSTRATUS. *Linn. Syft. Nat.* 677. 20.—*Fn. Suec.* 823.

CYCHRYS ROSTRATUS. *Payk. Fn. Suec.*

━━━━━━

Taken by G. Milne, Efq. F. L. S. The fpecies has been found in Scotland and Devenfhire, and though generally efteemed rare, was lately met with in confidérable plenty on mountains in Ireland by Mr. W. Leach, F. L. S.

LINNÆAN

LINNÆAN INDEX

TO

VOL. XIV.

COLEOPTERA.

HEMIPTERA.

N 2

LEPI-

INDEX.

LEPIDOPTERA.

NEUROPTERA.

HYMENOPTERA.

Bembex

INDEX.

DIPTERA.

APTERA.

ALPHA-

ALPHABETICAL INDEX

то

VOL. XIV.

INDEX.

Law and Gilbert, Printers, St. John's-Square, London.

THE

NATURAL HISTORY

OF

BRITISH INSECTS.

THE

NATURAL HISTORY

OF

BRITISH INSECTS;

EXPLAINING THEM

IN THEIR SEVERAL STATES,

WITH THE PERIODS OF THEIR TRANSFORMATIONS,
THEIR FOOD, ŒCONOMY, &c.

TOGETHER WITH THE

HISTORY OF SUCH MINUTE INSECTS

AS REQUIRE INVESTIGATION BY THE MICROSCOPE.

THE WHOLE ILLUSTRATED BY

COLOURED FIGURES,

DESIGNED AND EXECUTED FROM LIVING SPECIMENS.

BY E. DONOVAN, F.L.S.; W.S., &c.

VOL. XV.

LONDON:

PRINTED FOR THE AUTHOR,
And for F. C. and J. RIVINGTON, Nº 62. PAUL's CHURCH-YARD.

MDCCCXI.

505

THE

NATURAL HISTORY

of

BRITISH INSECTS.

━━━━━━

PLATE DV.

PHALÆNA TYPICOIDES.

SCARCE GOTHIC MOTH.

LEPIDOPTERA.

GENERIC CHARACTER.

Antennæ taper from the bafe : tongue fpiral : wings in general deflected when at reft.   Fly by night.

PHALÆNA TYPICOIDES. Wings incumbent, varied fufcous
and reddifh, with white ftreaks : pofterior pair
whitifh with fufcous border.

――――――

This is an infect of large fize, and elegant appearance, and is
generally efteemed among collectors in Britain on account of its
rarity.   It neverthelefs occurs fometimes in abundance in particular
fituations, as for example in fome parts of Yorkfhire, and alfo in
Devonfhire.   In the latter County at Knowle near Kingsbridge Mr.
Montagu takes it in confiderable plenty in his own garden throughout
the fummer.   The fame fpecies inhabits Germany.

The transformations of this curious Moth are altogether unknown.

PLATE DVI.

CARABUS GLABRATUS.

SMOOTH CARABUS.

COLEOPTERA.

GENERIC CHARACTER.

Antennæ filiform : feelers generally fix, the laft joint obtufe and truncated ; thorax flat and margined : wing-cafes margined.

SPECIFIC CHARACTER.

Apterous, black : wing-cafes very fmooth, and uniformly black.

CARABUS GLABRATUS. Apterus, ater elytris laviffimis unicoloribus. *Fabr. Ent. Syft. T.* 2. 125. 4.—Carabus gla-bratus *Panz.*

═══════

A rare fpecies defcribed by Fabricius as a native of Germany. It was firft difcovered in this country by Mr. W. G. Hooker of Norwich, who found it in great plenty on a mountain in Yorkfhire ; and fince that period it has been met with in the rocks near Killarney in Ireland.

B 2   PLATE

PLATE DVII.

MUSCA MORTUORUM.

DIPTERA.

GENERIC CHARACTER.

Mouth with a foft exferted flefhy probofcis and two equal lips: fucker furnifhed with briftles: feelers two, very fhort (or none) antennæ fhort.

SPECIFIC CHARACTER

AND

SYNONYMS.

Antennæ feathered: thorax black: abdomen braffy green: legs black.

MUSCA MORTUORUM: antennis plumatis, thorace nigro, abdomine viridi æneo, pedibus nigris. *Linn. Syft. Nat.* 2. 986, 66.—*Fn. Sv.* 1830.—*Fabr. Ent. Syft.* *T.* 4. 318. *n.* 23.

A native of Sweden, and other parts of Europe; it is to be efteemed alfo a britifh fpecies, a fpecimen having been lately taken

by

by Mr. W. E. Leach, in Scotland: this was met with on the bog
in which the battle of Bannockbourn was fought, a wafte in the
vicinity of Stirling, and is the fole authority within our recollection
upon which we could venture to confider it a native of thefe
kingdoms. It is a moft interefting acquifition, and is reprefented in
its natural fize with all poffible fidelity, in two different pofitions.

In addition to the character affigned to this fpecies by Linnæus, we
ought to obferve that the thorax is lineated with black, and greyifh, a
circumftance to which Fabricius refers in the expreffion " *thorax fub-
lineatus ;*" in the fpecimen above defcribed, thefe lines are pretty
diftinct.

P L A T E

508

PLATE DVIII.

FIG. I. I.

ELATER SANGUINEUS.

SANGUINEOUS ELATER, OR SKIPPER BEETLE.

COLEOPTERA.

GENERIC CHARACTER.

Antennæ filiform, lodged in a groove under the head and thorax : under-fide of the thorax terminating in an elaftic fpine, placed in a cavity of the abdomen, by which means when placed on the back it fprings up, and recovers its natural pofition.

SPECIFIC CHARACTER

AND

SYNONYMS.

Deep black : wing·cafes fanguineous, and without fpots.

ELATER SANGUINEUS: ater elytris ftriatis fanguineis immaculatis. *Linn. Syft. Nat.* 2. 654. 21.—*Fn. Sv.* 731.— *Gmel.* 1906. 21.
Elater nigris elytris rubris,—Le taupin à étuis rouges. *Geoff. Inf.* *I.* 131. 2.

Elater

Elater melanocephalus ruber *Panz. Voet. ii.* 117. 21. *t.* 44. *f.* 21.
ELATER SANGUINEUS. *Marſh. Ent. Brit. T. I. p.* 383. *n.* 20.

———————

Very rare in Britain. The ſpecies is found in Sweden, Germany and France.

———————

FIG. II II.

ELATER CUPREUS.

COPPERY ELATER, or SKIPPER BEETLE.

SPECIFIC CHARACTER
AND
SYNONYMS.

Coppery: wing-caſes half yellow: antennæ of the male pectinated.

ELATER CUPREUS: cupreus, elytris dimidiato-flavis: antennis maris pectinatis. *Marſh. Ent. Syſt. T. I. p.* 384. 23.
ELATER CUPREUS. *Fabr. Syſt. Ent.* 15.—*Spec. Inſ.* 1. 268. 20.
—*Ent. Syſt.* 1. *b.* 225. 37.
ELATER CASTANEUS. *Scop.* 286.

———————

A very beautiful and rare ſpecies: its ſize equal to that of Elater pectinicornis: dull coppery or braſſy rufous; antennæ black.

PLATE

2

1

PLATE DIX.

CURCULIO SULCIROSTRIS.

SULCATED-SNOUT CURCULIO, or WEEVIL BEETLE.

COLEOPTERA.

GENERIC CHARACTER.

Antennæ clavated, and feated on the fnout which is horny and prominent: feelers four and filiform.

SPECIFIC CHARACTER.

Oblong, cinereous, clouded with black: fnout with three furrows.

CURCULIO SULCIROSTRIS: oblongus cinereus fubnebulofus roftro trifulcato. *Fabr. Inf.* 1. *p.* 187. *n.* 143.—*Mant. Inf.* 1. *p.* 114. *n.* 185.

CURCULIO SULCIROSTRIS. *Gmel. Linn. Syft. Nat.* 1787. 85.— *Paykull Monogr.* 100.—*Panz. Ent. Germ.* 321. 128.—*Marfh. Ent. Brit. T.* 1. *p.* 308. *n.* 204.

———

Found on plants in Europe; it is a fpecies of large fize, and is rare in Britain.

VOL. XV.　　　　　　C　　　　　　PLATE

PLATE DX.

FIG. I.

PHALÆNA OMICRONATA.

O-O CARPET MOTH.

LEPIDOPTERA.

GENERIC CHARACTER.

Antennæ gradually tapering from the bafe : tongue fpiral : wings in general defle&ed when at reſt. Fly by night.

* *Geometra.*

SPECIFIC CHARACTER.

PHALÆNA OMICRONATA. Firſt wings fulvous at the bafe and tip, with fufcous ſtreaks and dots : in the middle a broad indented cinereous band with a double O nearer the anterior margin.

An elegant and not very frequent ſpecies found in the vicinity of woods in the month of June, July and Auguſt.

c 2

FIG.

FIG. II.

PHALÆNA CONTRISTATA.

COMMON CARPET MOTH.

SPECIFIC CHARACTER.

PHALÆNA CONTRISTATA. White with a common fufcous border :
  anterior wings brownifh at the bafe, the middle
  with a fufcous band : difk of the pofterior pair
  with dufky ftreaks.
PHALÆNA CONTRISTATA. *Marfh. M. S.*

————

Common in the months of May and June in moft woody
fituations. It is very abundant, in particular near London.

PLATE

PLATE DXI.

MUSCA FESTIVA.

DIPTERA.

GENERIC CHARACTER.

Mouth with a soft exserted fleshy probofcis, and two equal lips: sucker furnished with briftles: feelers two, very short (or none) antennæ short.

SPECIFIC CHARACTER.

MUSCA FESTIVA. Green and golden, gloffed with red, and cinereous: thorax lineated with yellow: abdomen banded with black.

––––––––––

An insect of peculiar beauty, extremely rare, and which does not appear to be defcribed by any author.

The natural fize of this very curious fpecies is denoted by the fmaller figure in the upper part of the plate: when magnified, as is fhewn beneath, its appearance is remarkable for its fingularity. The head is rather fmall and green, with eyes of a deep fufcous colour: the thorax green gloffed with blue, with a double golden yellow line in

the

the middle, and a fingle one on each fide : the abdomen greenifh, and partaking more of the rich metallic hues of gold, yellow, and tints of red, than the other parts. The whole of the body is of a fomewhat flender form; the wings large, and the legs long and flender in proportion

PLATE

PLATE DXII.

CURCULIO CARDUI.

THISTLE CURCULIO, OR WEEVIL BEETLE.

COLEOPTERA.

GENERIC CHARACTER.

Antennæ clavated, and feated on the fnout which is horny and prominent: feelers four and filiform.

SPECIFIC CHARACTER.

CURCULIO CARDUI. Oblong: green, black with numerous dots and broken linear bands of yellowifh down: wing-cafes ftriated with impreffed dots.

———

A new infect lately difcovered by Mr. W. E. Leach, from whom it received the fpecific appellation of Cardui: it occurs on the thiftle, but is rare.

PLATE

PLATE DXIII.

ATTELABUS MELANUROS.

COLEOPTERA.

GENERIC CHARACTER.

Antennæ moniliform, thicker towards the tip, and feated on the fnout: head pointed behind and inclined.

SPECIFIC CHARACTER

AND

SYNONYMS.

Black: wing-cafes teftaceous, the tip black.

ATTELABUS MELANUROS: niger, elytris teftaceis apice nigris. *Gmel.*
Linn. Syft. Nat. 1810. 6.
ODACANTHA MELANURA. *Paykull. Fn. Sv. I.* 169.—*Fabr. Syft.*
Eleut. 1. 228.—*Latr. Gen. Inf. I.* 194.—*Tab.*
meth. 164.
Carabe retréce. *Oliv. Entom.* 3. 35.

This is a fpecies rather exceeding, in point of fize, fuch infects as may with propriety be termed diminutive; its length, as fhewn by

VOL. XV. D the

the fmalleft figure, exceeding one third of **an inch,** and including the antennæ confiderably more: its form is peculiar, and the colours which are gay, in fome degree remarkable for their brilliancy. Its singularity confifts in the very curious form of the thorax, a kind of elongated cylinder, connecting the head with the body, as if the former were placed on a flender pedicle; the thorax being narrower by one half than the head, and not above one third the breadth of the wing-cafes. Notwithftanding this difproportion of its parts, the appearance is not devoid of elegance, and to this the beauty of its colours contribute materially.

The head and thorax of this infect are green and blue, changeable into each other, and highly gloffy: the contraft between the colours of thefe and the wing-cafes is ftriking, the latter being fine orange, with the exception of a large common fpot of the fame fhining blue and green, as on the head and thorax, or rather inclining more to azure, that is, difpofed at the pofterior extremity. It is no lefs worthy of remark, that the lower furface is in like manner varied with blueifh fhining green, and orange, the head, thorax and pofterior part of the abdomen, being of the former colour, and **the** intervening fpace of the abdomen, orange. The antennæ are orange from the bafe to the middle, beyond which they are dufky: the legs alfo are of two colours, the thigh and firft joint being orange, the remainder dufky; and befides this the thighs are black at the tips.

Gmelin defcribes this fpecies as a native of Upfal. In Britain it is a very local fpecies, but does not appear uncommon in the places it inhabits: it occurs abundantly in Cromllyn bog, in Gla-morganfhire, near Swanfea, and alfo in a bog near Norwich.

PLATE

514

PLATE DXIV.

PHALÆNA CONVERSARIA.

LARGE BANDED CARPET MOTH,

LEPIDOPTERA.

GENERIC CHARACTER.

Antennæ taper from the bafe: tongue fpiral: wings in general deflected when at reft. Fly by night.

Geometra.

SPECIFIC CHARACTER

AND

SYNONYMS.

Wings pale brown, with dark fufcous middle band, and greyifh indented common band behind: pofterior legs deeply ciliated.

GEOMETRA CONVERSARIA. *Hübn. Schmet.* 62. 321 ?

D 2   This

This interefting acquifition to the Britifh Entomologift was difcovered by Mr. W. E. Leach, the latter end of Auguft, about the
year 1807, in Warley-wood, at Tamerton, near Plymouth, Devonfhire.

The fize of this infect is confiderable, the colour above pale brown,
with a rich dark fufcous band acrofs the middle, and immediately
behind it, a broad and very pale common band, circumfcribed above
by an angulated, and beneath by an indented palifh line. The whole
furface is fprinkled with fpecks of brown. The under furface is
paler, with more obfolete fpeckling, and fome blotches of fufcous,
forming an interrupted common band in the middle. The four
anterior legs are naked as ufual, the two pofterior deeply fringed with
fine hairs.

PLATE

PLATE DXV.

DERMESTES MURINUS.

COLEOPTERA.

GENERIC CHARACTER.

Antennæ clavated, the club perfoliated, and three of the joints thicker: thorax convex and flightly margined: head inflected, and concealed under the thorax.

SPECIFIC CHARACTER
AND
SYNONYMS.

Oblong, black, clouded with whitifh down: abdomen and breaft white.

DERMESTES MURINUS : oblongus tomentofus nigro alboque, abdomine niveo *Linn. Syft. Nat.* 2. 156. 3. 18.— *Fn. Sv.* 426.

DERMESTES MURINUS. *Fabr. Ent. Syft. T. I. p. I.* 230. 14.

DERMESTES MURINUS : tomentofus fufco cinereoque nebulofus, fcutello fulvo. *Marfh. Ent. Brit. T. I. p.* 61. 2.

DERMESTES NEBULOSUS. *De Geer. Inf.* 4. 197. 2.

Feeds on putrid carcafes.

PLATE

PLATE DXVI.

CARABUS BIPUSTULATUS.

BIPUSTULATED CARABUS.

COLEOPTERA.

GENERIC CHARACTER.

Antennæ filiform: feelers generally fix, the laft joint obtufe and truncated ; thorax flat and margined : wing-cafes margined.

SPECIFIC CHARACTER

AND

SYNONYMS.

Winged : thorax orbicular, and with the anterior part of the wing-cafes rufous, pofterior part black, with a common rufous fpot.

CARABUS BIPUSTULATUS : alatus, thorace orbiculato rufo, coleoptris apice nigris : macula rufa. *Fabr. Ent. Syft.*
T. I. 151. *n.* 164.—*Paykull. Fn. Sv. I.* 138.
54.—*Marfh. Ent. Brit. T. I. p.* 464. 88.

———

A fmall, but elegant fpecies, and which appears to peculiar advantage when magnified : the ground colour varies from rufous to
paler,

paler, yellowifh, and teftaceous in different fpecimens; and fome little variation is perceptible alfo in the form of the common fpot at the pofterior part of the wing cafes. The antennæ are fufcous at the bafe, the extremity pale or yellowifh, and the legs of the latter colour.

The fmalleft figure reprefents this pretty little infect in its natural fize.

PLATE

517

PLATE DXVII.

PHALÆNA INSCRIPTATA.

LETTERED MOTH.

LEPIDOPTERA.

GENERIC CHARACTER.

Antennæ taper from the bafe : tongue fpiral : wings in general deflected when at reft. Fly by night.

* * * *Geometra.*

SPECIFIC CHARACTER.

PHALÆNA INSCRIPTATA. Pale, anterior wings banded and lineated : with two dufky charaćters, and a whitifh Λ in the difk of the middle band : pofterior wings with fcalloped lines.

———————

The two moths reprefeuted in this plate, are beyond difpute, varieties of the fame fpecies, that delineated in the upper part of the plate differing only in having the anterior wings and bands darker than the other : the charaćters on both are the fame, being two fmall dufky letter-like marks, one of which remotely refembles the hebrew *kametz* (ד) and a little behind thefe is a pale or whitifh greek Λ (lambda.) The whole of thefe marks are difpofed near the center of the broad pale band that paffes acrofs the middle of the anterior wings, and thefe conftitute the principal character of the fpecies. In the darkeft coloured fpecimen of this infect, there is a fmall and pretty diftinct dot in the middle of the pofterior wings.

This is a very rare and apparently undefcribed fpecies of the Geo-metra tribe.

PLATE

PLATE DXVIII.

ELATER RUFICOLLIS.

RUFOUS-NECKED ELATER, or SPRINGER BEETLE.

COLEOPTERA.

GENERIC CHARACTER.

Antennæ filiform and lodged in a groove under the head and thorax: fides of the thorax terminated in an elaftic fpine placed in a cavity of the abdomen, by means of which the infect, when on its back, recovers its natural pofition.

SPECIFIC CHARACTER
AND
SYNONYMS.

Black and polifhed, anterior part of the thorax red

ELATER RUFICOLLIS: niger, thorace pofterius rubro nitido. *Linn. Fn. Suec.* 724.—*Fabr. Sp. Inf. I. p.* 270. *n.* 33 —*Mant. inf. I. p.* 173. *n.* 37.—*Ent. Syft. I. b.* 227. 52.

ELATER RUFICOLLIS: thorace rubro nitido antice nigro, elytris corporeque nigris. *Marfh. Ent. Brit. T. I.* 376. *n.* 2.

A rare and very pretty fpecies. Its habits are unknown. Linnæus defcribes it as a native of Sweden.

E 2   PLATE

PLATE DXIX.

FIG. I. II.

MUSCA ULIGINOSA.

DIPTERA.

GENERIC CHARACTER.

Mouth with a foft exferted flefhy probofcis, and two equal lips: fuckers furnifhed with briftles: feelers two, very fhort, or none: antennæ generally fhort.

* *Nemotelus*

Sucker with a fingle recurved briftle without fheath : feelers none : antennæ moniliform, the tip fetaceous, and inferted at the bafe of the probofcis.

SPECIFIC CHARACTER

AND

SYNONYMS.

Male black: abdomen whitifh, with black bands at the tip. *Female* dark with a dorfal line of pale fpots on the abdomen.

FIG.

PLATE DXX.

CHRYSOMELA LONGIPES.

LONG-LEGGED CHRYSOMELA.

COLEOPTERA.

GENERIC CHARACTER.

Antennæ moniliform: feelers fix, larger towards the end: thorax marginate: wing-cafes immarginate: body generally oval.

SPECIFIC CHARACTER.

Chrysomela longipes. Oblong: black: wing-cafes orange with a black fpot at the bafe, and two near the middle: anterior legs of the male very long.

———

A very curious infect taken about the month of May, by W. E. Leach, Efq. near Sidmouth, in Devonfhire.

Fig. I. I. exemplifies the upper and lower furface of the male infect, which is diftinguifhed by the difproportionate length of the anterior legs: fig. II. the female in which the anterior legs are about the fame length as the others. Both are reprefented in their natural fize.

VOL. XV.   F

PLATE DXXI.

ACRYDIUM SUBULATUM.

SUBULATE ACRYDIUM.

HEMIPTERA.

GENERIC CHARACTER.

Antennæ filiform, and inferted under the eyes: feelers filiform and equal: lip ovate, and cleft at the tip: head ovate and inferted: thorax carinated: fcutel produced behind to the end of the abdomen, and covering the wings: wing-cafes none, or only lamina: legs fhort: pofterior pair long, and formed for leaping, the tarfi compofed of three joints.

SPECIFIC CHARACTER
AND
SYNONYMS.

Dark brown: fcutel longer than the abdomen.

Acrydium subulatum: thoracis fcutello abdomine longiore *Fabr. Ent. Syft T. 2. 26.*
Gryllus subulatus. *Linn. Syft. Nat.* 2. 693. 8.—*Fn. Succ.* 865.

———

An interefting, and very curious little fpecies, and which is alfo rare in Britain. The fmaller figure denotes the natural fize.

F 2   PLATE

PLATE DXXII,

ICHNEUMON PERSUASORIUS.

HYMENOPTERA.

GENERIC CHARACTER.

Mouth with a ſtraight horny membranaceous bifid jaw, the tip rounded, and ciliated : mandibles curved and ſharp : lip cylindrical, membranaceous at the tip, and emarginate ; feelers four unequal and filiform, and ſeated in the middle of the lip: antennæ ſetaceous, of more than thirty joints: ſting exſerted, incloſed in a cylindrical ſheath, compoſed of two valves, and not pungent.

SPECIFIC CHARACTER

AND

SYNONYMS.

Scutel white : thorax ſpotted : all the ſegments of the abdomen with two white dots on each ſide.

ICHNEUMON PERSUASORIUS : ſcutello albo, thorace maculato, abdominis ſegmentis omnibus utrinque punctis duobus albis *Linn. Syſt. Nat.* 2. 932. 16.—*Fn. Sv.* 1593.—*Fabr. Ent. Syſt. T.* 2. 145. *n.* 49.

This

This curious ſpecies is a native of the North of Europe, and Germany. Linnæus deſcribes it as a Swediſh inſect; Panzer and Schaeffer as an inhabitant of Germany; and Walckenær as being found in the environs of Paris. In Britain it is very rare: we have only heard of three Britiſh ſpecimens, one of which was taken by Mr. W. J. Hooker of Norwich.—Its transformations are not deſcribed by any writer.

The figures in the annexed plate repreſent the ſpecies in its natural ſize.

PLATE

PLATE DXXIII

LIBELLULA SCOTICA.

SCOTCH LIBELLULA.

NEUROPTERA.

GENERIC CHARACTER.

Mouth armed with more than two jaws: lip trifid: antennæ very thin, filiform, and fhorter than the thorax: wings expanded: tail of the male furnifhed with a forked procefs.

SPECIFIC CHARACTER.

LIBELLULA SCOTICA. Thorax with two oblique yellow bands.

Male. Wings tranfparent with deep black ftigma: abdomen blackifh.
Female. Wings tranfparent with deep black ftigma, and yellow bafe: abdomen yellow, with two black lines on each fegment.

———

We have been recently favoured with fpecimens of this new fpecies of Libellula by W. E. Leach, Efq. from whom it received the trivial name of Scotica, in reference to the country in which it appears only to have been hitherto difcovered. This gentleman informs us it is common in the bogs of Scotland: he firft obferved it near Lock-awe,

in

in Argylefhire, and afterwards in the bog of Bannock-bourn, in which latter place it occurs in great abundance.

Libellula Scotica is an infect of the middle fize, in general appearance refembling the fpecies *vulgata*. The male is uniformly dufky, except the wings, which are tranfparent : the female is more remarkable for its gaiety, the head, thorax, and abdomen being yellowifh, varied with brown, and little lines of black; and the wings tranfparent, with the bafe yellow.

PLATI

PLATE DXXIV.

CURCULIO BITÆNIATUS.

COLEOPTERA.

GENERIC CHARACTER.

Antennæ clavated, and feated on the fnout which is horny and prominent: feelers four and filiform.

SPECIFIC CHARACTER

AND

SYNONYMS.

Thorax brown, with a pale ftreak each fide: wing-cafes cinereous fprinkled with fufcous, ftriated with impreffed dots, and marked between the ftriæ with numerous blackifh points.

CURCULIO BITÆNIATUS: thorace fufco: linea utrinque pallida, elytris cinereis fufco-confperfis punctato-ftriatis. *Marfh. Ent. Brit. T. I. p.* 268. 93 *.

---

* Roftrum nigricans, thorace brevius. Thorax fufcus, ex ovato rotundus, linea utrin-que pallida. Elytra fufco cinerea, ftriata; ftriæ punctis impreffis: inter ftrias punctula plurima nigricantia. *Ib.*

This remarkable fpecies of Curculio appears to be defcribed only by Mr. Marfham, to whofe fpecific and general defcription nothing material can be added. Its habitat was not apparently known till after the publication of *Entomologia Britannica,* as that work is filent in this particular: it is now afcertained that the fpecies is not uncommon among grafs, in the winged ftate, and that the grafs affords the larva its natural, and favorite food. The neft or reticulated open work cone which it conftructs previoufly to becoming a pupa, and in which it remains while in that ftate, is extremely delicate and curious in its fabric, though not in thefe refpects fingular, for many of the Curculiones form fimilar cafes in which they remain enveloped while in the ftate of pupa. The pupa cafe of Curculio bitæniatus, is reprefented in the annexed plate.

PLATE

525

PLATE DXXV.

HISTER QUADRIMACULATUS.

FOUR-SPOTTED HISTER.

COLEOPTERA.

GENERIC CHARACTER.

Antennæ clavated, the club folid: the laft joint compreffed, and curved: head retractile within the body: mouth forcipated: wing-cafes fhorter than the body, and truncated: anterior fhanks denticulated, the hind fhanks fpinous.

SPECIFIC CHARACTER

AND

SYNONYMS.

Black: wing-cafes fub-ftriated, with two red fpots.

HISTER QUADRIMACULATUS: ater, elytris fubftriatis, maculis duabus rubris. *Linn. Syft. Nat.* 567. 6.—*Fn. Su.* 443.—*Paykull. Fn. Suec.* 36. 2.—*Marfh. T. I.* p. 94. *fig.* 6.

G 2   A rare

A fcarce Britifh fpecies, found in the dung of animals : it alfo in-habits Germany and other parts of Europe.

Its fize furpaffes that of Hifter unicolor; the colour black, with a large lunated fpot of red on each of the wing-cafes. Sometimes thefe fpots are interrupted in the middle, and in fuch fpecimens the wing-cafes exhibit the four diftinct red fpots which the fpecific name implies. The lower furface is entirely black, and both the upper and lower furfaces are remarkably gloffy.

The appearance of this infect, when the wing-cafes and wings are expanded, is very fingular : this is reprefented in the annexed plate, together with its afpect in a quiefcent ftate.

PLATE

PLATE DXXVI.

CARABUS INTRICATUS.

INTRICATE-DOTTED CARABUS.

COLEOPTERA.

GENERIC CHARACTER.

Antennæ filiform : feelers generally fix, the laft joint obtufe and truncated ; thorax flat and margined : wing-cafes margined.

SPECIFIC CHARACTER

AND

SYNONYMS.

Apterous, violet-black : wing-cafes with raifed intricate ftriæ and dots.

CARABUS INTRICATUS.   Apterus violaceo-niger, elytris intricatis elevato-ftriatis punctulatifque. *Linn. Fn. Suec.* 780. &c.

CARABUS CYANEUS.   Apterus niger violaceo nitens, elytris punctis intricatis rugofis *Paykull. Monogr.* 10. 2 — *Fabr. Ent. Syft. T. I. p.* 126. *n.* 9.—Bupreftis nigroviolaceus. *Geoffr. Inf. I.* 144. 4.

It

It appears Fabricius was aware the carabus he defcribed under the fpecific name of cyaneus, muft be in all refpects the fame as the Linnæan Carabus intricatus, fince he inferts the reference to that fpecies in *Fauna Suecica,* among his fynonyms : Paykull called it cyaneus, and this name Fabricius was induced to retain, though certainly lefs applicable than that it had previoufly obtained from Linnæus.—Fabricius defcribes it as a native of woods in Europe. Panzer includes it among the infects of Germany : in England it is very rare, and indeed appears not to have been difcovered in the latter country till very lately.

The figure reprefents this curious infect in its natural fize.

PLATE

527

PLATE DXXVII.

PHALÆNA ROBORARIA.

GREAT OAK BEAUTY MOTH.

LEPIDOPTERA.

GENERIC CHARACTER.

Antennæ taper from the bafe: tongue fpiral: wings in general deflected when at reft. Fly by night.

Geometra.

SPECIFIC CHARACTER

AND

SYNONYMS.

Wings indented, grey with numerous brown ftreaks and fpecks: beneath whitifh, lower wings, with a fufcous lunule in the middle.

PHALÆNA ROBORARIA: pectinicornis alis dentatis grifeis: atomis ftrigifque numerofis fufcis. *Fabr. Ent. Syft. T.* 3. *p.* 2. 137. *n.* 28.—*Efp. T.* 5. *f.* 2.

Several

Several very beautiful, and rather diſtinct varieties of this fine ſpecies occur in Auſtria : that which we have delineated is the only variety, however, we believe found in England, where it appears to be extremely uncommon.   The ſpecies we apprehend to be rare on the Continent, as well as in England, Fabricius referring expreſly to the cabinet of Mr. Scieffermyller, for the example he deſcribes.

Phalæna roboraria is nearly allied to the ſpecies of geometra, deno-minated by Engliſh collectors the " mottled beauty," (Phalæna repan-daria) from which it is diſtinguiſhed by the ſuperiority of its ſize, and ſome little variation in the form and diſpoſition of the fuſcous lines on the wings : the diſſimilarity is evident on an accurate compariſon, but is not ſo obvious at the firſt view.

Fabricius deſcribes the larva as being of a grey brown colour, with a darker dorſal line, and curves on the ſegments, and according alſo to this writer the larva feeds on the oak, whence it obtains the trivial appellation of the great oak beauty.

P L A T E

528

PLATE DXXVIII.

FIG. I. I.

CANTHARIS FASCIATA.

FASCIATED CANTHARIS.

COLEOPTERA.

GENERIC CHARACTER.

Antennæ filiform: thorax generally margined, and fhorter than the head: wing-cafes flexile: fides of the abdomen edged with folded papillæ.

SPECIFIC CHARACTER

AND

SYNONYMS.

Thorax greenifh: wing-cafes blackifh, with two red bands.

CANTHARIS FASCIATA. *Linn. Syft. Nat.* 648. 10.—*Fn. Sv.* 711. —*Gmel.* 1899. 10.

VOL. XV.   H   MALA.

Malachius fasciatus: elytris nigris: fasciis duabus rufis. *Fab.*
     Syst. Ent. 208. 4.—*Sp. Inf. I.* 262. 5.—*Mant.*
     I. 169. 8.—*Ent. Syst. I. a.* 224. 13.
Malachius fasciatus.   *Oliv. Inf.* 27. 10. 12. *tab. I. fig.* 2.
Cantharis fasciata: thorace virefcente, elytris nigris: fasciis
     duabus rubris.   *Marsh. Ent. Brit. T.* 371. 11.
Telephorus fasciatus. *De Geer.* 4. 76. 9.
La cicidele à bandes rouges.   *Geoff. I.* 177. 12.

     This is a very gay and pretty insect: the antennæ and legs are
black: the head blackish, glossed with shining green, as is likewife
the thorax: the wing-cafes are dusky purple with a broad band of
red acrofs the middle, and another behind formed by the junction
of the tips of the wing-cafes, the latter being of the fame red colour
as the band in the middle.   The abdomen at the fides are red.

     The fmalleft figure denotes the natural fize.   This fpecies is found
among mofs.

FIG.

FIG. II.

CANTHARIS BIPUSTULATA.

BIPUSTULATED CANTHARIS.

SPECIFIC CHARACTER.

Braſſy green: front yellowiſh : wing-caſes red at the tip.

CANTHARIS BIPUSTULATA: aeneo-viridis, fronte flavicante elytris apice rubris. *Marſh. Ent. Brit. T. I. p.* 369. 9.

CANTHARIS BIPUSTULATA. Aeneo-viridis, elytris apice rubris. *Linn. It. oel.* 127.—*Fn. Suec.* 709.—*Gmel. Linn. Syſt. Nat.* 1893. 8.

MALACHIUS BIPUSTULATUS. *Oliv. Inſ.* 27. 5. 3. *t. I. f. I.—Fabr. Syſt. Ent.* 208. 2.—*Sp. Inſ. I.* 262. 2. *Mant. I.* 169. 2.—*Ent. Syſt. i. a.* 222. 2.

TELEPHORUS BIPUSTULATUS. *De Geer.* 4. 75. 7.

DONACIA ASPARAGORUM. *Panz. Voel. ii. th.* 128. 6. *t.* 46. *f.* 6.

La cicindele verte á points rouges. *Geoff. Inſ. I.* 175. 8.

———————

Frequent among graſs: the larva rapacious, feeding on ſmaller inſects, and even the grubs of its own tribe and ſpecies.

H2   **PLATE**

PLATE DXXIX.

CURCULIO PINI.

PINE CURCULIO.

COLEOPTERA.

GENERIC CHARACTER.

Antennæ clavated and feated on the fnout, which is horny and prominent.

SPECIFIC CHARACTER
AND
SYNONYMS.

Black, wing-cafes fufcous, with dots and clouded bands of yellowifh.

CURCULIO PINI: niger, elytris fufcis: fafciis nebulofis. *Linn. Syft. Nat.* 608. *n.* 19.—*Fn. Suec.* 589.—*Gmel.* 1746. 19.—*Marfh. Ent. Brit. T. I. p.* 289. 152.

A rare fpecies in England, but common in Scotland upon the *Pinus fylveftris*. According to Linnæus, it is found in Sweden ; and Panzer includes it among the infects of Germany.

The fmalleft figure denotes the natural fize of this curious infect.

PLATE

PLATE DXXX,

CARABUS LUNATUS.

LUNATED CARABUS.

COLEOPTERA.

GENERIC CHARACTER.

Antennæ filiform: feelers generally fix, the laft joint obtufe and truncated; thorax flat and margined: wing-cafes margined.

SPECIFIC CHARACTER
AND
SYNONYMS.

Thorax orbicular and rufous: wing-cafes yellow, with three black fpots.

CARABUS LUNATUS. Thorace orbiculato rufo, elytris flavis: maculis tribus nigris. *Fabr. Ent. Syft. I.* 163. 172. —*Syft. Ent.* 247. 60.—*Panz. Ent. Germ.* 63. 98.—*Marfh. Ent. Brit. T. I. p.* 466. 1.
Carabus eques. *Schranck. Beytr.*

———————————

This very elegant little fpecies is defcribed by Fabricius as an inhabitant of Britain, on the authority of a fpecimen in the collection

of

of Mr. Lee: the species occurs likewife in Germany, and Italy. Whether it is common in thefe latter mentioned countries is uncertain, we fufpect not: in Britain it is rare.

The appearance of this infect when magnified is interefting; the natural fize is denoted by the fmalleft figure.

P L A T E

PLATE DXXXI.

SPHEX SPIRIFEX.

HYMENOPTERA.

GENERIC CHARACTER.

Mouth with an entire jaw: mandibles horny, incurved, and denticulated: lip horny and membranaceous at the tip: feelers four: antennæ with ten articulations: wings in each fex incumbent and flat: fting pungent and concealed within the abdomen.

SPECIFIC CHARACTER

AND

SYNONYMS.

Black: thorax hairy, immaculate: petiole of one joint, yellow, and as long as the abdomen.

SPHEX SPIRIFEX: atra thorace hirto immaculato, petiolo uniarticulato flavo longitudine abdominis. *Fabr. Ent. Syft. T. 2. 204. 24.—Schæff. Icon. tab. 38. fig. I.*

———————

We poffefs an example of this curious infect in the Britifh cabinet of the late Mr. Drury. The fpecimen does not exactly feem to accord with the Linnæan Sphex Spirifex, but rather with the ac-

VOL. XV.   I   knowledged

knowledged variety of that fpecies defcribed by Linnæus under the name of Ægyptia ; and is clearly the variety found by Schæffer in the environs of Ratifbon (Ichneumon *decimus feptimus*) to which Fabricius refers for his fpecies fpirifex.

This infect is chiefly an inhabitant of the fouth of Europe, where it lives in focieties: the nefts are conftructed in the fides of the mud-walls of cottages and other convenient fituations; their form cylindrical, and the interior in fome degree refembling a honey-comb. It preys on infects of every kind, and is in particular a great enemy to the fpider, which it eafily overcomes, and feems to prefer to moft other food.—The fize of this fpecies is confiderable.

PLATE

PLATE DXXXII.

FIG. I. I.

STAPHYLINUS LUNULATUS.

LUNULATED ROVE-BEETLE.

COLEOPTERA.

GENERIC CHARACTER.

Antennæ moniliform: feelers four: wing-cafes half as long as the body: wings folded up under the wing-cafes: tail armed with a forceps, and furnifhed with two exfertile veficles.

SPECIFIC CHARACTER

AND

SYNONYMS.

Thorax and abdomen orange: wing-cafes black with two orange fublunate fpots at the bafe: extremity of the abdomen black, with a pale band.

STAPHYLINUS LUNULATUS: rufus, capite abdominis elytrorumque pofticis nigris, femoribus totis rufis. *Linn. Fn. Sv.* 845.—*Gmel.* 2037. 7.—*Paykull. Monogr.* 41.

I 2   OXYPORUS

OXYPORUS LUNULATUS: flavus elytris nigris bafi apiceque pallidis.
Fabr. Syft. Ent. 268. 2.—*Sp. Inf. i.* 338. 2.—
Mant. i. 219. ♀.—*Ent. Syft. I. b.* 532. 3.
STAPHYLINUS LUNULATUS. *Marfh. Ent. Brit. T. I.* 523. 72.

———————

A minute and very beautiful fpecies, found in the dung of cattle.—
The natural fize of this, and the other two interefting little fpecies
reprefented in the annexed plate, is denoted by the fmaller figures.
Staphylinus lunulatus is a native of the northern parts of Europe,
being found in Sweden and Denmark, as well as Britain, and extends
likewife as far fouthward as Germany and France.

———————

FIG. II. II.

STAPHYLINUS MARGINATUS.

MARGINATED ROVE-BEETLE.

SPECIFIC CHARACTER

AND

SYNONYMS.

Black: fides of the thorax, two dots on the anterior part of the
wing-cafes, with the pofterior margin, and the legs rufous.

STAPHYLINUS

Staphylinus marginatus: ater thoracis lateribus pedibufque flavis. *Fabr. Syſt. Ent.* 266. 8.—*Sp. Inſ. i.* 336. 9.—*Mant. i.* 22. 15.—*Ent. Syſt. i. b.* 526. 30.
Paykull. Monogr. 32.—*Fn. Suec. iii.* 392. 32. *Gmel. Syſt. Nat* 2028. 36.

Oxyporus marginellus. *Panz. Ent. Germ.* 355. 21.

Staphylinus marginatus: ater, thoracis lateribus pedibufque rufis. *Marſh. Ent. Brit.* 512. 40.

Fabricius defcribes this ſpecies as a native of England and Norway.

FIG. III. III.

STAPHYLINUS BIPUSTULATUS.

BIPUSTULATED ROVE-BEETLE.

SPECIFIC CHARACTER

AND

SYNONYMS.

Black : wing-cafes with two ferruginous dots.

Staphylinus bipustulatus: niger, elytris puncto ferrugineo. *Marſh. Ent. Brit. T. I.* 527.

Staphylinus

STAPHYLINUS BIPUSTULATUS. *Fabr. Syſt. Ent.* 266. 11.—*Sp.*
 Inſ. i. 336. 12.—*Mant. i.* 221. 18.—*Ent. Syſt.*
 i. b. 526. 34.
OXYPORUS BIPUSTULATUS. *Panz. Faun. Germ.* 27. *t.* 10.

———————

Inhabits various parts of Europe.

PLATE

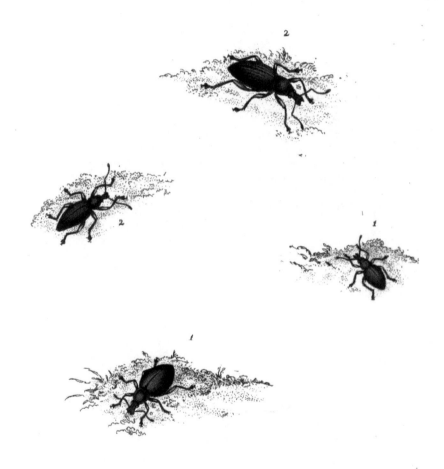

PLATE DXXXIII.

CURCULIO MARITIMUS.

MARITIME CURCULIO.

COLEOPTERA.

GENERIC CHARACTER.

Antennæ clevated, and feated on the fnout which is horny and prominent: feelers four and filiform.

SPECIFIC CHARACTER.

CURCULIO MARITIMUS. Body fomewhat ovate, black, rather gloffy and minutely punctured, with obfolete ftriæ on the wing-cafes.

———————

An interefting infect very nearly allied to Curculio niger of Ento-mologia Britannica, but is ftill fufficiently diftinct, and fhould moft probably be confidered as a new fpecies. It was found by W. E. Leach, Efq. on *Arenaria maritima and peploides*, and might hence have borne the fignificant appellative of *arenarius*, had not that name been previoufly given by Herbft to a very different infect difcovered in Pruffia, and which is alfo retained by Gmelin under the fame name in *Syftema Naturæ*.

A material

A material difference is perceptible in the two infects at prefent under confideration, notwithftanding which, they are, however, pre-fumed to conftitute the two fexes of the fame fpecies. One is fmaller and more ovate than the other, the ftriæ of punctures on the wing-cafes lefs obvioufly defined, and the thighs blackifh, while in the other the thighs incline to ferruginous red. The fmaller delineations at fig. I. and II. exemplify the natural fize a little enlarged, the other reprefentations exhibit the fame infects confiderably magnified.

PLATE

PLATE DXXXIV.

LYTTA VESICATORIA.

BLISTER-BEETLE.

COLEOPTERA.

GENERIC CHARACTER.

Antennæ filiform: head gibbous, inflected, and broader than the thorax: thorax in general cylindrical: wing-cafes foft, flexile, and linear: body elongated.

SPECIFIC CHARACTER

AND

SYNONYMS.

Bright gloffy golden green: antennæ black.

LYTTA VESICATORIA: viridiffima nitens, antennis nigris. *Marfh. Ent. Brit. T. I.* 484. *n. I.—Fabr. Sp. Inf. I. p.* 328. *n.—I. Mant. Inf. I. p.* 215. *n. I.—Syft. Ent.* 260. *I.—Ent. Syft. I. b.* 83. *I.*

MELOE VESICATORIUS. *Linn. Syft. Nat.* 679. 3. *Fn. Suec.* 827.

Cantharis veficatoria De Geer. 5. 12. *t. I. f.* 9.

La cautharide des boutiques. *Geoff. I.* 341. 1. *t.* 6. *f.* 5.

The Lytta veficatoria is a very uncommon infeɛt in this country though abundant in many other parts of Europe, the ſouthern eſpecially; and is in particular found in ſuch plenty in Spain as to have obtained from that circumſtance the popular appellation of Spaniſh Cantharis, or bliſter-fly.

Mr. Marſham defcribes it as a Britiſh ſpecies, and we poſſeſs a ſpecimen in the cabinet of the late Mr. Drury, which is admitted to have been taken alive in this country. We have beſides another ſpecimen which differs only from the former in being entirely of a blue inſtead of green colour, and this we are induced to repreſent likewiſe from the perfuaſion of its being a Britiſh infeɛt. In a natural ſtate the Lytta veficatoria is found on the privet and the elder, and alſo on the aſh; its uſes in the pharmacy are ſufficiently underſtood, and theſe alone, independently of its beauty, muſt doubtleſs entitle this elegant little infeɛt to our immediate confideration.

P L A T E

PLATE DXXXV.

FIG. I.

ELATER CYANEUS.

CYANEOUS SKIPPER-BEETLE.

COLEOPTERA.

GENERIC CHARACTER.

Antennæ filiform, lodged in a groove beneath the head and thorax : under-fide of the thorax terminating in an elaftic fpine, lodged in a cavity of the abdomen, by which means when placed on the back it fprings up, and recovers its natural pofition.

SPECIFIC CHARACTER

AND

SYNONYMS.

Entirely purplifh blue and dotted : wing-cafes ftriated.

ELATER CYANEUS : totus purpureo-caeruleus punctulatus, elytris ftriatis. *Marfh. Ent. Brit. T. I.* 388. 32.

━━━━━━━━━━

A rare infect, and which appears hitherto to have been defcribed only as a Britifh fpecies. Some collectors have conceived it to be a fexual difference of the following kind.

κ 2   FIG.

FIG. II.

ELATER IMPRESSUS.

IMPRESSED SKIPPER-BEETLE.

SPECIFIC CHARACTER

AND

SYNONYMS.

Blue-black, and glofsy : wing-cafes with an impreffed dorfal line, and two dots : legs pitchy.

ELATER IMPRESSUS : atro-cæruleus niditus, thorace linea dorfali punctifque duobus impreffis, pedibus piceis. *Marfh. Ent. Brit. T. I.* 387. 29.

Rather larger than the preceding.

PLATE

PLATE DXXXVI.

BOMBYLIUS MINOR.

SMALL HUMBLE BEE FLY.

DIPTERA.

GENERIC CHARACTER.

Mouth with a very long ſtraight ſetaceous ſucker, formed of two unequal horizontal valves, and containing ſetaceous ſtings.

SPECIFIC CHARACTER
AND
SYNONYMS.

Wings immaculate: body yellowiſh, hairy: legs teſtaceous.

BOMBYLIUS MINOR: alis immaculatis, corpore flaveſcente hirto, pedibus teſtaceis. *Linn. Syſt. Nat.* 1009. 4.— *Fn. Sv.* 1920.—*Fabr. Ent. Syſt. T.* 4. 409. *n.* 9.

———

This is the laſt of the Bombylius tribe we have to deſcribe in the preſent work; three ſpecies only being found in Britain and of theſe two are already included.

The

The three fpecies of Britifh Bombylii bear the names of major, medius, and minor; and thefe are fufficiently expreffive of the comparative fize of each, major being the largeft, minor lefs by one half; and medius of an intermediate fize between the other two. The latter appears to be more rare than either: it is found in fpring, hovering over flowers the nectareous juices of which, afford its favourite food. The figure reprefents this little fpecies in its natural fize.

PLATE

537

PLATE DXXXVII.

FIG. I.

SILPHA HUMATOR.

COLEOPTERA.

GENERIC CHARACTER.

Antennæ clavated, the club perfoliated: wing-cafes margined: head prominent: thorax fomewhat flattened and margined.

SPECIFIC CHARACTER

AND

SYNONYMS.

Oblong, entirely black, except the rufous tip of the antennæ.

SILPHA HUMATOR: oblonga tota atra, antennis apice rufis exceptis.
Marſh. Ent. Brit. T. I. 114. 2.
NICROPHORUS HUMATOR. *Olivier. Inſ.* 2. 10. 8. 4. *tab. I. fig.* 2.
—*Fabr. Ent. Syſt. T. I. p.* 247. *n.* 2.
DERMESTES. *Geoff. Inſ. I.* 99. 2.

———

Rare in Britain. This infect is fimilar to the fpecies Germanica, but differs in being fmaller, and in having the whole of the clavated

part

part of the antennæ except the firſt joint rufous. The prevailing colour is black inclining to chocolate ; the head, thorax, wing-cafes, extremity of the abdomen and legs, and alfo the whole of the under furface being of this colour. That part of the abdomen above which is covered by the wings and wing-cafes when the infeᴄt is at reſt is teſtaceous, and the tip of the abdomen, or tail, the fame but rather darker.

FIG. II.

SILPHA MORTUORUM.

SPECIFIC CHARACTER

AND

SYNONYMS.

Oblong, black ; wing-cafes with a band and ſpot of ferruginous : club of the antennæ black.

SILPHA MORTUORUM : oblonga atra, elytris faſcia maculaque fer-
             rugineis. *Marſh. Ent. Brit. T. I.* 115. 4.
NICROPHORUS MORTUORUM. *Fabr. Ent. Syſt. I. a.* 248. 5.
NICROPHORUS VESPILLOIDES. *Fueſl. Archiv.* 89. *I.*
Pollinᴄtor vulgaris minor. *Voet. Coleopt. t.* 30. 3.

Similar to Silpha veſpillo, from which it differs in the following material particulars : it is ſmaller than veſpillo : the clavated part of the antennæ is black inſtead of ferruginous : the rufous orange ſpaces

on

on the wing-cafes inftead of being difpofed in two diftinct tranfverfe
bands form a fingle common band acrofs the middle, and a detached
fpot behind on each of the wing-cafes.   Sometimes the anterior
band is in like manner divided by the furrounding fpace of black into
two diftinct reddifh fpots, one on each wing-cafe as in the pofterior
part before defcribed, and by that means exhibits altogether four
orange fpots, two on each fide.—It may be laftly added, that the legs
are naked in the fpecies mortuorum, while in S. Vefpillo thefe are
befet with fulvous down.

This infect feeds on carrion and dung, like the other fpecies to
which it is clofely allied : Panzer has alfo found it in Fungi.

538

PLATE DXXXVIII.

BOLETARIA MULTIPUNCTATA.

COLEOPTERA.

GENERIC CHARACTER.

Antennæ perfoliated, and thicker towards the end: thorax margined, with three hollows behind, the middle one obsolete: wing-cases margined: body ovate.

SPECIFIC CHARACTER

AND

SYNONYMS.

Black: wing-cases finely striated, varied with black and ferruginous.

BOLETARIA MULTIPUNCTATA: nigra, elytris minutè striatis, ferrugineo nigroque variis. *Marsh. Ent. Brit. T. I. p.* 139. 3.

MYCETOPHAGUS MULTIPUNCTATUS. *Fabr. Ent. Syst. i. b.* 498. 5. *Panz. Ent. Germ.* 337. 4.

DERMESTES MULTIPUNCTATUS. *Thunb. Inf. Suec.* 79. 6.

————

Found on fungi, of the Boletus genus. Length two lines.

L 2   FIG.

FIG. II. II.

BOLETARIA ATOMARIA.

SPECIFIC CHARACTER

AND

SYNONYMS.

Black, with dots and band behind fulvous.

BOLETARIA ATOMARIA: nigra, elytris punctis fasciaque poftica
         fulvis. *Marfh. Ent. Brit. T. I. p.* 141. *n.* 7.
IPS ATOMARIA. *Fabr. Mant. I.* 46. 9.
Mycetophagus atomarius. *Fabr. Ent. Syft. i. b.* 498. 4.

———

.This is about the fize of the former: the general colour black
gloffed with blue : the head and thorax immaculate : wing-cafes varied
with dots, and an irregular waved pofterior line : antennæ and legs
dufky.

It is found like the laft, and fucceeding fpecies on fungi of the
Boletus kind.

FIG.

FIG. III. III.

BOLETARIA PUNCTATUS.

SPECIFIC CHARACTER

AND

SYNONYMS.

Pitchy: wing-cafes fomewhat punctated and black, band at the bafe and two fpots at the end of the wing-cafes ferruginous.

MYCETOPHAGUS PUNCTATUS : piceus elytris fubpunctatis nigris bafi ferrugineis. *Fabr. Ent. Syft. T. I. b.* 499. *n.* 10. *Panz. Ent. Germ.*

———

Rare, and rather larger than the two preceding.

PLATE

539

PLATE DXXXIX.

FIG. I. II.

SILPHA SINUATA.

SINUATE SILPHA.

COLEOPTERA.

GENERIC CHARACTER.

Antennæ clavated, the club perfoliated: wing-cafes margined: head prominent: thorax fomewhat flattened and margined.

SPECIFIC CHARACTER

AND

SYNONYMS.

Thorax emarginate and very rough: wing-cafes with three raifed lines, and the tip finuate.

SILPHA SINUATA: thorace emarginate rugofiflimo: elytris lineis elevatis tribus apice finuatis. *Marfh. Ent. Brit. T. I.* 120. 14.

SILPHA SINUATA. *Fabr. Syft. Ent.* 75. 13.—*Sp. Inf. I.* 88. 16. —*Ent. Syft. I. a.* 252. 18.—*Gmel.* 1622. 56.

Le Bouclier noir a corcelet raboteux. *Geoff. I.* 119. 2.

This

This is a fpecies of moderate fize, allied in habit to Silpha obfcura. The prevailing colour is blackifh, inclining to grey, dull and without glofs : the thorax is brownifh and rugofe, with a filky hue and fome-what filvery. The two fexes are diftinguifhed by the termination of the wing-cafes, this in one being much finuated, and forming a diftinct lobe, the other nearly entire.—This laft mentioned infect is the Silpha opaca of fome writers.

PLATE

540

PLATE DXL.

DYSTISCUS PUNCTULATUS.

DOTTED BOAT-BEETLE.

COLEOPTERA.

GENERIC CHARACTER.

Antennæ fetaceous: feelers fix, filiform: hind-legs formed for fwimming, fringed on the inner fide, and nearly unarmed with claws.

SPECIFIC CHARACTER

AND

SYNONYMS.

Black: wing-cafes with three rows of dots: fhield of the head, margin of the thorax, and wing-cafes yellow.

DYTISCUS PUNCTULATUS. niger clypeo thoracis elytrorumque margine albis, elytris ftriis tribus punctatis. *Geoffr. Inf. I.* 185. *I.*—Le Ditique brun à bordure. *ib.*

DYTISCUS PUNCTULATUS. *Fabr. Ent. Syft. I. a.* 188. 4.

VOL. XV.   M   *DYTISCUS*

DYTISCUS LATERALI-MARGINATUS. *Degeer. T. 4. p.* 396. *n.* 3.

DYTISCUS VIRENS. *Müll. zool dan. prodr. p.* 70. *n.* 664.

DYTISCUS PUNCTATUS. *Oliv.* 3. 40. 22. 4. *t. I. f.* 6. *b. and f. I. e.*

DYTISCUS PUNCTULATUS. *Marſh. Ent. Brit. T. I. p.* 412. *n.* 2.

This in common with the other ſpecies of the Dytiſcus genus is found in marſhes, ponds, and other waters, particularly thoſe of the ſtagnant kind, and which abound moſt with the refuſe of animal ſubſtances, and aquatic plants.

The larva, like the reſt of its tribe, is active, fierce, and vigorous; entirely aquatic and ſubſiſts on the other ſmall inhabitants of the regions in which it lives, ſuch as the larva of the Ephemeræ, the Phry-ganæ and many other creatures of the infect race that ſpend the earlier ſtage of their exiſtence in the watery element; and alſo on the vermes which in ſuch ſituations occur invariably, and in conſiderable numbers. In their turn the larvæ of the Dytiſci become the food of fiſhes, and aquatic birds, but rarely fall a prey to the infect race, as their natural ſtrength, and the powerful armament of their jaws, at leaſt in the larger ſpecies, enables them to maintain a decided ſuperiority over theſe puny enemies.

Previous to its paſſing into the pupa ſtate, the larva of this ſpecies emerges from the bottom of the water, and forms a convenient receptacle for the purpoſe in ſome adjacent bank, or ſpot of ground near the water's edge; this accompliſhed, it changes to the pupa, and after a while appears in the winged ſtate.—Having aſſumed this

form

form it becomes in all refpects an amphibious creature, refiding alternately in the water, or on the land. When in the water, however, which appears to be its moft congenial element, it is frequently obferved to rife upon the furface to take in air, and on the contrary when on land, or in flight, it does not willingly remain a long time before it again plunges into the aquatic element.

Dytifcus punctulatus is found in the ditches of Batterfea meadows.

m 2

LINNÆAN

LINNÆAN INDEX

TO

VOL. XV.

COLEOPTERA.

INDEX.

HEMIPTERA.

LEPIDOPTERA.

NEUROPTERA.

HYMENOPTERA.

INDEX.

HYMENOPTERA.

DIPTERA.

ALPHA-

ALPHABETICAL INDEX

VOL. XV.

━━━━━━━

N

melanuros,

INDEX.

Law and Gilbert, Printers, St. John's Square, London.

THE

NATURAL HISTORY

OF

BRITISH INSECTS.

Law and Gilbert, Printers, St. John's Square, London.

THE

NATURAL HISTORY

OF

BRITISH INSECTS;

EXPLAINING THEM

IN THEIR SEVERAL STATES,

WITH THE PERIODS OF THEIR TRANSFORMATIONS,
THEIR FOOD, ŒCONOMY, &c.

TOGETHER WITH THE

HISTORY OF SUCH MINUTE INSECTS

AS REQUIRE INVESTIGATION BY THE MICROSCOPE.

THE WHOLE ILLUSTRATED BY

COLOURED FIGURES,

DESIGNED AND EXECUTED FROM LIVING SPECIMENS.

BY E. DONOVAN, F.L.S. W.S., &c.

VOL. XVI.

LONDON:

PRINTED FOR THE AUTHOR,
And for F. C. and J. RIVINGTON, Nº 62, ST. PAUL'S CHURCH YARD.

MDCCCXIII.

541

THE

NATURAL HISTORY

OF

BRITISH INSECTS.

———

PLATE DXLI.

PAPILIO ARTAXERXES.

ARTAXERXES BUTTERFLY.

LEPIDOPTERA.

GENERIC CHARACTER.

Antennæ clavated at the tip: wings erect when at reft: fly by day.

SPECIFIC CHACACTER

AND

SYNONYMS.

Wings entire, black, with a white dot in the middle of the anterior pair, and rufous lunules on the pofterior ones: margin beneath white, with rufous dots.

VOL. XVI.   B   PAPILIO

PAPILIO ARTAXERXES.   *Jon. M. S.—Pict.* 6. *tab.* 44. *fig.* 2.
HESPERIA ARTAXERXES : alis integerrimis nigris : anticis puncto
        medio albo, posticis rufis, fubtus margine albo
        rufo punctato.   *Fabr. Ent. Syft. T.* 3. *p.* 1. 297.
        129.
LYCÆNA ARTAXERXES.   *Fabr. Syft. Gloffat.*

———————

To the great aftonifhment of our Englifh Collectors of Natural
Hiftory in the vicinity of the metropolis, Papilio Artaxerxes, an infect
heretofore efteemed of the higheft poffible rarity, has been lately
found in no very inconfiderable plenty in Britain : for this interefting
difcovery we are indebted to the fortunate refearches of our young and
very worthy friend, W. E. Leach, Efq. who met with it common on
Arthur's Seat near Edinburgh, and alfo on the Pentland Hills.

A difcovery fo interefting in the annals of Entomology deferves
efpecial notice, because Papilio Artaxerxes was not merely efteemed
rare in this country ; on the continent it appears to be totally unknown : their Entomologifts, till the time of Fabricius, have not mentioned it, nor had Fabricius himfelf ever feen an example of the
fpecies ; he derived his information folely from a drawing by the hand
of W. Jones, Efq. of Chelfea.   The extreme accuracy of that delineation, it muft be indeed allowed, would render it unneceffary for
Fabricius to confult the infect from which it was pourtrayed, but the
circumftance is mentioned in order to prove the rarity of the fpecies
as an European infect ; and we cannot, it is prefumed, afford a more
decifive teftimony of its intereft in this refpect than in ftating Fabricius,
its original defcriber, had never feen it.

Papilio Artaxerxes is by no means striking in appearance ; it becomes important from the general eftimation of its fcarcity, and its claim
to confideration in this view is indubitable.   In the beft of the Englifh
                                     cabinets,

cabinets, with the exception of that of our fincere friend A. M'Leay, Efq. we have often lamented to obferve a deception intended to fupply the deficiency of this fpecies; namely, a little painting of the infeft, carefully configned on a pin, to the moft obfcure corner of the drawer, amongft the Britifh Papiliones, and which, from its fpecious afpeft and ingenious fimilitude, has oftentimes, we fufpeft, been miftaken for the original: this is a general fault, arifing undoubtedly from a very pardonable motive, and therefore, we apprehend, fhould not be re-prehended in terms of unufual feverity; yet we cannot think the cuftom wholly blamelefs.—We have alluded to the cabinet of Mr. M'Leay, and it will be therefore right to add in explanation, that his valuable and extenfive collection contained a very fine fpecimen of **Papilio Artaxerxes**, that had been taken in Scotland previoufly to the difcovery made by Mr. Leach, as before related.

Though we are not difpofed to concede this little Butterfly any confiderable portion of praife on account of its beauty, it is not alto-gether devoid of claim in this refpeft: the upper furface differs little from feveral analogous fpecies, the females of feveral of " the blues," as they are ufually termed, at the fame time that the afpeft of the lower furface is entirely diffimilar from moft others; and exhibits a very delicate, fpotted, and prettily diverfified appearance.

As thefe infefts fly in the day-time there can be little doubt they may be fought for by the Collector with fuccefs on the hilly fpot called Arthur's Seat, near Edinburgh.

u 2   PLATE

542

PLATE DXLII.

COCCINELLA 4-PUNCTATA.

FOUR-SPOT RED LADY-COW.

COLEOPETERA.

Antennæ clavated, club folid: anterior feelers femicordated: thorax and wing-cafes margined: body hemifpherical: abdomen beneath black.

SPECIFIC CHARACTER

AND

SYNONYMS.

Wing-cafes red with four black dots.

COCCINELLA 4-PUNCTATA: coleopteris flavis: punctis nigris quatuor. *Linn. Syft. Nat.* 2. 580. 590. *Fabr. Sp.
Inf.* 1. *p.* 95. *n.* 16.—*Mant. Inf.*
1. *p.* 56. *n.* 28.
COCCINELLA QUADRIPUNCTATA. *Gmel. Linn. Syft. Nat*
1647. 9.
COCCINELLA 4-PUNCTATA. *Marfh. Ent. Brit. T.* 1. *p.* 151. 7.

Profeffor Gmelin, in defcribing this fpecies of Coccinella, refers for its name and character exclufively to the Fabrician *Species Infectorum,* and *Mantiffa,* and hence it might be concluded that Fabricius was its

firft

firſt defcriber, which is not the fact, as it was previouſly noticed in the
Linnæan Syſtema Naturæ; this overfight would be deemed of fome
importance in any work, but muſt be of ſtill greater moment in a production profeſſedly defigned as an improved edition of the Linnæan
publication: nor is the circumftance the lefs remarkable, fince Fabricius, in the work quoted, affords a reference to the defcription previouſly given by Linnæus.

Linnæus, and alfo Fabricius, fpeak of it in general terms as an inhabitant of Europe: the fpecies does not, however, appear, to be by
any means frequent like many others of the fame genus; for, with the
exception of the works of Villers, and thofe abovementioned, it does
not occur in any of the continental publications on Entomology. The
author of *Entomologia Britannica*, T. Marſham, Efq. introduces
this fpecies for the firſt time to notice as a Britifh infect: the fpecimen
he defcribes is in the cabinet of Dr. Shaw. Another was taken in the
town of Plymouth on the 18th of September, 1812, and is now in
the poſſeſſion of W. E. Leach, Efq.—The reader will pardon the
minutenefs of this detail when the rarity of the fpecies is duly
eftimated.

In its general afpect this uncommon infect differs little from feveral
others of the fame tribe that are very abundant, and in which the
wing-cafes are red, with dots of black, and the thorax yellow with an
irregular fpot or fplafh of black in the middle: it is alfo like thofe of
the middle fize; the body beneath is black.

P L A T E

PLATE DXLIII.

PHALÆNA MARGARITARIA.

LIGHT EMERALD MOTH.

LEPIDOPTERA.

GENERIC CHARACTER.

Antennæ taper from the bafe: wings in general deflected when at reft. Fly by night.

SPECIFIC CHARACTER

AND

SYNONYMS.

Wings angular, whitifh green with a deeper band terminating in a white ftreak.

PHALÆNA MARGARITARIA. *Linn. Syft. Nat.* 5. 865. 231.

PHALÆNA MARGARITARIA: alis angulatis albidis: faturatiore ftriga alba terminata. *Fab. Ent. Syft.* 3. 131. 10.

GEOMETRA MARGARITARIA. *Hübn. Schmett Geom.* 3. 13.

━━━━━━━━━━

Inhabits England and Germany in woods, and feeds principally on the Carpinus and Betula. The Moth appears in July and Auguft Its larva is defcribed: the form is elongated, with two white dots on the laft fegments: tail bifid, and feet twelve in number.

Both fexes of this Moth are reprefented in the annexed plate; that with the antennæ larger or more deeply pectinated is the male.

PLATE

PLATE DXLIV.

LAMPYRIS FESTIVA.

FESTIVE LAMPYRIS.

COLEOPTERA.

GENERIC CHARACTER.

Antennæ filiform: feelers four: wing-cafes flexile: thorax flat, femi-orbicular, furrounding and concealing the head; fegments of the abdomen terminating in folded papillæ: female ufually apterous.

SPECIFIC CHARACTER.

Lampyris Festiva. Sublinear, tawny orange: wing-cases with four raifed lines, and pofterior end black: disk of the thorax black.

———————

An infect of more ftriking afpect, notwithftanding the inferiority of its fize, can fcarcely prefent itfelf. It appears to be extremely rare; indeed, we have not feen it in any other cabinet than that of the late Mr. Drury, now in our own poffeffion; nor does it feem to be defcribed by any author; we prefume, therefore, it may be new to Entomologifts in general.

VOL. XVI.　　　　　C　　　　　Ia

In the Linnæan Syſtem this inſect muſt fall under ſome one of the ſeveral families into which modern Naturalists divide the Lampyrides of that author. Neverthelefs, it ſhould not be concealed that its characters are in certain refpects remote from that of the true Lampyrides, and might, without any degree of impropriety, remove it entirely from that genus. The Entomologiſt need ſcarcely to be apprized of the families to which we allude; the genus Lampyris, as eſtabliſhed by Linnæus, it muſt be known, are divided into many ſections, without which it would be impoſſible to retain the whole of the Lampyrides under one generic appellation. Even Gmelin, in editing the laſt edition of Syſtema Naturæ, ſeems ſenſible of this, as he forms no lefs than five diſtinct families for their reception. The whole of theſe, according to Fabricius, from whom they are adopted, are generically diſtinct, and are ſo conſtituted by him under the refpective names *Lampyris, Omalyſſus, Coſſyphus, Pyrochroa,* and *Lycus:*—this is the order in which they ſtand in the lateſt works of that writer: ſome further alterations have been again made by writers ſubfequent to Fabricius, but generally, it may be obſerved, the example of Fabricius is almoſt implicitly followed in the diviſion of the Linnæan genus of Lampyrides, throughout the continent of Europe.

That the preſent inſect is a genuine Lampyris of Linnæus admits of no doubt; but in referring it to either of the ſections, we ſhould exprefs ourſelves with greater caution; for perhaps it ought rather to conſtitute a diſtinct genus than be configned to either. It has the characters of Pyrochroa, and yet is allied to Lycus. As a fpecies it ſeems to approach the Pyrochroa, called by Herbſt *Aurora,* a native of Pomerania; but ſhould it be the ſame, there muſt be a deficiency in the character aſſigned to it by that author, which creates uncertainty; neither the black ſpot in the diſk of the thorax, nor thoſe at the extremity of the wingcafes, being there deſcribed: we conclude for theſe reafons it cannot be the ſame.

The highly beautiful form of the pectinated antennæ in this elegant little inſect contributes to render its general appearance attractive:
the

the form is graceful, and the colours pleasing,—a fine tawny orange
diverfified with characteriftic marks and fpots of black. The antennæ are brown except the extreme joint, which is tawny: on the
upper furface it will be obferved, that in the centri part of the
olack difk of the thorax is a raifed acute line, and on the wing-cafes
four diftinctly prominent ftriæ, with the interftices deeply punctured.
It is perceptible beneath that all the thighs at the bafe are tawny.

c 2   P L A T E

PLATE DXLV.

ELATER 4-PUSTULATUS.

FOUR-SPOT SPRINGER BEETLE.

COLEOPTERA.

GENERIC CHARACTER.

Antennæ filiform, lodged in a groove under the head and thorax: under fide of the thorax terminating in an elaftic fpine, placed in a cavity of the abdomen; by which means the body, when placed on the back, fprings up and recovers its natural pofture.

SPECIFIC CHARACTER

AND

SYNONYMS.

Black wing-cafes ftriated, with two teftaceous dots.

ELATER 4-PUSTULATUS: niger elytris ftriatis: punctis duobus teftaceis. *Fabr. Ent. Syft. T.* 1. *p.* 2. *p.* 235. *fp.* 89.—*Paykull,* &c.

—————

A diminutive infect, of very uncommon rarity, originally defcribed by Fabricius from a fpecimen in the cabinet of Hybner: this was found in Saxony: it has been fince defcribed as a native of Sweden, and was lately discovered on the banks of the Tavy river.

The fmalleft figure denotes the natural fize; the prevailing colour is black, the fpots on the wing-cafes with the legs teftaceous.

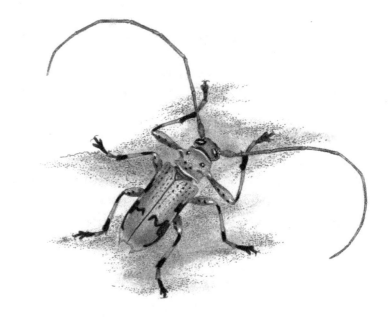

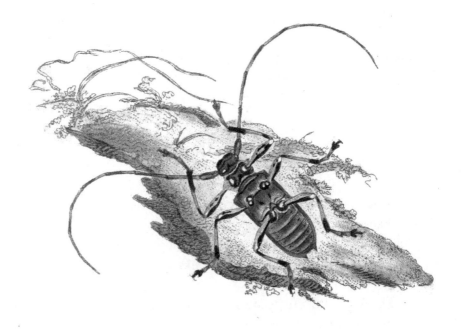

PLATE DXLVI.

CERAMBYX LITERATUS.

LETTERED CERAMBYX.

COLEOPTERA.

GENERIC CHARACTER.

Antennæ fetaceous: feelers four: thorax fpinous or gibbous: wing-cafes linear.

SPECIFIC CHARACTER.

CERAMBYX LITERATUS. Cinereous: bafe of the wing-cafes with raifed black dots: exterior margin brown, with two confluent fpots, the posterior one branching into a letter-form band.

———————

A new Britifh fpecies of the Cerambyx tribe, lately found in the vicinity of Manchester by John King, Efq. and kindly communicated to the Author. It is an infect of confiderable fize, in appearance remarkable for its elegance, and does not appear to have been defcribed or figured in any other publication.

PLATE

PLATE DXLVII.

FIG. I.

SCARABÆUS VERNALIS.

SPRING SCARABÆUS.

COLEOPTERA.

GENERIC CHARACTER.

Antennæ clavated, the club fiffile: fhanks of the anterior legs generally dentated.

SPECIFIC CHARACTER

AND

SYNONYMS.

Wing-cafes glabrous and very fmooth; fhield of the head rhomboidal: crown flightly prominent.

SCARABÆUS VERNALIS: elytris glabris læviffimis, capitis clypeo rhombeo: vertice prominulo. *Linn. Syft. Nat.* 551. 43.—*Fn. Suec.* 389.—*Paykull Fn. Suec.* 1. 6. 6.—*Fabr. Syft. Ent.* 17. 61.—*Sp. Inf.* 1. 19. 75.—*Mant.* 1. 10. 82.—*Ent. Syst.* 1. *a.* 31. 98.—*Gmel. Linn. Syst. Nat.* 1549. 43.—*Geoffr.* 1. 77. 10.—*Fourc.* 1. 7. 10,—*Herbft. Arch.* 1. 7. 19.

VOL. XVI,   D

7. 19.—*Panz. Faun. Germ.* 49. *t.* 2.—*Sulz.*
Hift. t. 1. *f.* 6.—*Marfh. Ent. Brit. T.* 1. *p.* 23.
n. 37.
Geotrupes vernalis. *Latr. Gen. Cruft. et Inf. T.* 2. *p.* 94.

———

Allied to Scarabæus ftercorarius, from which it differs chiefly in being fmaller, and in having the wing-cafes fmooth inftead of being furrowed. There is a variety of this fpecies, in which the violefcence is not very perceptible, and which has been occafionally miftaken, on that account, for a diftinct fpecies.

———

FIG. II.

SCARABÆUS SYLVATICUS.

SYLVAN SCARABÆUS.

SPECIFIC CHARACTER

AND

SYNONYMS.

Gloffy violaceus: thorax at each fide impreffed: wing-cafes fome-what ftriated: feet pitchy.

Scarabæus sylvaticus: violaceus nitidus, thorace utrinque im-preffo, elytris fubftriatis, tarfis piceis. *Marfh.*
Ent.

Ent. Brit. T. 1. *p.* 23. 38.—*Paykull. Fn. Suec.*
1. 55.—*Panz. Ent. Germ.* 8. 31.—*Scriba Ephem.*
3. 250.

GEOTRUPES SYLVATICUS. *Latr. Gen. Cruft. et Inf. Vol.* 2.
p. 93.

About the fize of the former.   Found in dung.

D 2 PLATE

548

PLATE DXLVIII.

FIG. I. I.

PHRYGANEA MONTANA.

MOUNTAIN SPRING-FLY.

NEUROPTERA.

GENERIC CHARACTER.

Mouth with a horny short curved mandible: feelers four: stemmata three: antennæ setaceous, longer than the thorax: wings equal, incumbent, and the lower ones folded.

SPECIFIC CHARACTER

AND

SYNONYMS.

PHRYGANEA MONTANA. Anterior wings testaceous with daubs and confluents transverse marks of fuscous: posterior wings pale, border with alternate fuscous and pale spots.

———————

Found abundant on the borders of rocky mountain streams in Wales, and similar situations in Ireland and other parts of Britain. The smallest figure, as in the following species, denotes the natural size.

FIG.

FIG. II. II.

PHRYGANEA MACULATA.

SPOTTED SPRING-FLY.

SPECIFIC CHARACTER

AND

SYNONYMS.

PHRYGANEA MACULATA : Anterior wings pale teftaceous towards
the bafe, and faintly reticulated with dusky : dorfal edge with four diftinct dark fpots, and a
feries of dark dots next the border at the apex.

━━━━━━━━

A new fpecies, found in tolerable plenty on the rivers of Cumber-
land and Devonfhire.

PLATE

549

PLATE DXLIX.

MUSCA ATHERIX.

ATHERIXINE MUSCA.

DIPTERA.

GENERIC CHARACTER.

Mouth with a soft exserted fleshy proboscis, and two unequal lips: ucker beset with bristles: feelers short and two in number, or sometimes none: antennæ usually short.

* Antennæ moniliform, with a terminal bristle.

SPECIFIC CHARACTER

AND

SYNONYMS.

Black: body with a grey spot on each side of all the segments: in the middle of the costal margin of the wings a dusky spot surrounded by a crescent of hyaline dots, and a dusky spot at the base.

ATHERIX MACULATA (mas) *Meig. Clafs. und. Befch. t. 1. p. 274.*

This insect, which is clearly of the Musca tribe in the system of Linnæus, constitutes a new genus in the work of Meigel, under the name of Atherix.—The last-mentioned genus has been recently adopted by Latreille in his subdivisions of the Muscæ.

There

There is an appearance of novelty and fimple elegance in this little infect that ftrongly demands attention, though the colours are merely black and dusky, with a diverfity of the grey,—fuch in fact as in the more emphatic language of practical collectors might be called the widow's weeds, or half-mourning. The fpecies is very rare.

PLATE

PLATE DL.

OESTRUS OVIS.

SHEEP BOT-FLY, GAD-FLY, or BREEZE-FLY.

DIPTERA.

GENERIC CHARACTER.

Mouth with a fimple aperture, and not exferted: feelers two, each confifting of two articulations, with the tip orbicular, and feated on each fide in a depreffion of the mouth: antennæ of three joints, the laft fubglobular, and furnifhed at the anterior part with a briftle, placed in two hollows of the front.

SPECIFIC CHARACTER

AND

SYNONYMS,

OESTRUS OVIS. Wings tranfparent, with fpots at the bafe and dotted nerves: abdomen filky white, varied with black dots and fpots.
Reaum. Inf. t. 36. fig. 22.—Larva. 8, 9.

OESTRIS OVIS. Wings tranfparent, with fmall fpots at the bafe: abdomen chequered with black and white. *Clarke, Linn. Tranf. V. 3. p.* 313. *tab.* 23. *fig.* 14—17.

This is one of thofe deftructive creatures which infeft quadrupeds, and are known under the general appellation of the Bot-flies. The particular fpecies now before us is that peculiar to the Sheep tribe.

VOL. XVI.   E   Every

Every hufbandman is acquainted with the direful effects produced by
the ravages of thofe intruders in the vital economy of that ufeful race
of creatures, though few are correctly acquainted with their hiftory.
The Bots, indeed, have never been fufficiently or properly defcribed
till within the laft few years, when the fubject was inveftigated by
Bracy Clarke, Efq. and it muft be added, with a degree of accuracy
highly creditable to himfelf and fatisfactory to the public. The refult
of his remaiks appeared firft in a Memoir publifhed in the third
Volume of the Linnæan Tranfactions, and fubfequently in other publications.   Thefe obfervations may be truly faid to form the bafis of
our prefent knowledge of the Oeftri tribe, than which no race of
infects whatever, can be more ftrictly deferving the attention of the
Entomologift or the obfervation of the Agriculturift.

In defcribing the individual fpecies of Oeftrus at prefent under confideration Mr. Clarke obferves, that about the middle of June he
procured fome full-grown larvæ from the infide of the cavity of the
bone which fupports the horns of the Sheep. They were nearly as
big as thofe of the large Horfe-bot, of a delicate white colour, flat
on the under fide and convex on the upper, having no fpines at the
divifions of the fegments, though provided with two curved hooks at
the fmall end : the other extremity is truncated, with a fmall prominent
ring or margin, which feems to ferve the fame purpofe, though in an
inferior degree, as the lips of the Oeftrus equi and hæmorrhoidalis, by
occafionally clofing over and cleaning the horny plate of refpiration.

The larvæ are perfectly white and tranfparent when young, except
the horny plates, which are black: as they increafe in fize the fegments
of the upper fide become marked with two brown tranfverfe lines,
and fome fpots are obfervable at the fides. They move with confiderable quicknefs, holding with the tentacula as a fixed point, and
drawing up the body towards them. The under-fide of the body is
marked with a broad line of dots, which, on examination with glaffes,
appear to be rough points, ferving perhaps the double purpofe of
affifting their paffage over the fmooth and lubricated faces of thefe
                                                              membranes,

membranes, and of exciting alfo a degree of inflammation in them where they reft, fo as to caufe a fecretion of lymphor pus for their food.

Mr. Clarke obferves, that he has moftly found thefe animals in the horns and frontal finufes, though he has remarked that the membranes lining, thefe cavities were hardly at all inflamed, while thofe of the maxillary finufes were highly fo; and hence he was led to fufpect that they inhabit the maxillary finufes, and crawl, on the death of the animal, into those fituations in the horns and frontal finufes. The breeds, he prefumes, are not confined to any particular feafon, as the young and full-grown larvæ are found together at the fame time.

The larvæ, when full grown, fall through the noftrils to the ground, and change to the pupa ftate, lying on the earth or adhering by the fide to a blade of grafs: in this ftate it remains about two months, when the fly appears.—The manner in which this fpecies depofits its eggs is difficult to obferve, owing to the obfcure colour and rapid motions of the fly, and the extreme agitation of the fheep; but from the mode of defence the fheep takes to avoid it, and its manners afterwards, there can be little doubt that the eggs are depofited in the inner margin of the noftril.

The moment the fly touches the noftril of the fheep, the latter fhake their heads violently, and beat the ground with their feet, holding their nofes at the fame time clofe to the earth, and running away, earneftly looking on every fide to fee if the fly purfues: they alfo may fometimes be feen fmelling to the grafs as they go, left one fhould be lying in wait for them; which if they obferve, they gallop back, or take fome other direction, as they cannot, like horfes, take refuge in the water. To defend themfelves againft its attacks they have recourfe to a rut, or dry dufty road, or gravel-pits, where they crowd together during the heat of the day, with their nofes held clofe to the ground, which renders it difficult for the fly, who attacks on the wing, to get at the noftril.

E 2

Perhaps,

Perhaps, fays Mr. Clarke, (in concluding his general obfervations,) the removal of the fheep to a diftant pafture during the months
of June and July, whilft the greater number of the Bots are yet on
the ground in the ftate of pupa, and not bringing them again on fuch
ground till the fetting in of winter, would be the means of deftroying
them moft effectually; and this procefs, repeated for two or three
years fucceffively, in places where the Oeftri are particularly trouble-
fome, might prove eventually ufeful to the farmer.

A highly-magnified figure of this fpecies in the winged ftate is
fhewn in the annexed Plate, from which it will appear an infect of
fingular character, and no very inconfiderable beauty. The fmall
figure denotes the natural fize.

PLATE

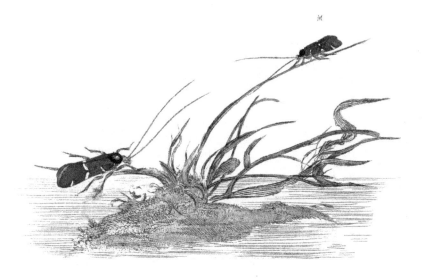

PLATE DLI.

PHRYGANEA INTERRUPTA.

BROKEN-BAR PHRYGANEA.

NEUROPTERA.

GENERIC CHARACTER.

Mouth with a horny short curved mandible: feelers four: stemmata three: antennæ setaceous, longer than the thorax: wings equal, incumbent, lower ones folded.

SPECIFIC CHARACTER

AND

SYNONYMS.

Wings black, with an interrupted white band of dots near the middle, and white dot on the anterior margin nearer the tip.

PHRYGANÆA INTERRUPTA: nigra alis nigris: fasciis quatuor niveis; anticis interruptis, posteriori marginali punctata. *Fabr. Ent. Syst. T. 2. p.* 79. 25.

———————

Common among willows on the banks of rivers and rivulets in various parts of Britain. Frequent about Plymouth, and on the banks of the Dee. Fabricius describes it as a native of England.

PLATE

PLATE DLII.

STAPHYLINUS HIRTUS.

HAIRY STAPHYLINUS, or ROVE-BEETLE.

COLEOPTERA.

GENERIC CHARACTER.

Antennæ moniliform: feelers four: wing-cafes half as long as the body: wings folded up under the wing-cafes: tail armed with a pair of forceps, and furnifhed with two exfertile veficles.

SPECIFIC CHARACTER.

AND

SYNONYMS.

Hairy, black: thorax and pofterior part of the abdomen yellow.

STAPHYLINUS HIRTUS: hirfutus niger, thorace abdomineque poftice flavis. *Linn. Syfl. Nat.* 683. 1.—*Fn. Suec.*
839.—*Gmel. Linn. Syfl.* 2025. 217. 1.
Fabr. fp. inf. 1. *p.* 334. *n.* 1.—*Mant. Inf.* 1. *p.*
219. *n.* 2. *Ent. Syfl.* 1. *b.* 519. 2.
Staphylinus niger villofus, &c.—Le Staphylin bourdon. *Geoffr. Inf.*
par. 1. *p.* 363. *n.* 7.
Staphylinus bombylius. *Degeer* 4. 20. 5.
STAPHYLINUS HIRTUS. *Marfh. Ent. Brit. T.* 1. *p.* 496. 1.

━━━━━━━━━

The largeft and moft interefting fpecies of the Staphylinus genus found in this country, and alfo one of the moft uncommon.

In

In Entomologia Britannica it ſtands recorded as a Britiſh ſpecies, and though esteemed rare, it uſually occurs in the beſt cabinets. It is ſaid to inhabit ſandy places, and is also found among moſs concealed or lying under ſtones. Mr. Comyns has met with it in Devonſhire. We once ſaw it on the wing in a thicket in Coombe Wood, Surrey.

PLATE

553

PLATE DLIII.

FIG. I. I.

CERAMBYX SANGUINEUS.

SANGUINEOUS CERAMBYX.

COLEOPTERA.

GENERIC CHARACTER.

Antennæ fetaceous: eyes lunate, and embracing the bafe of the antennæ: feelers four: thorax fpinous: wing cafes linear: body oblong.

SPECIFIC CHARACTER

AND

SYNONYMS.

Black: thorax fomewhat tuberculated, and with the wing-cafes fanguineous: antennæ moderate.

CERAMBYX SANGUINEUS: niger, thoracis dorfo elytrifque fanguineis, antennis mediocribus. *Marfh. Ent. Brit.* T. 1. 336. *n.* 19.
CERAMBYX SANGUINEUS. *Linn. Syft. Nat.* 636. 80.—*Fn. Suec.* 673.—*Gmel.* 1855. 80.

VOL. XVI.   F   CALLIDIUM

CALLIDIUM SANGUINEUM: thorace fubtuberculato elytrifque fanguineis, antennis mediocribus. *Fabr. Syft. Ent.* 190. 2.—*Sp. Inf.* 1. 238. 16.—*Mant. I.* 153. 25. *Ent. Syft.* 1. *b.* 326. 35.
Le Lepture veloutée couleur de feu. *Geoff.* 1. 220. 21.

———

This very beautiful fpecies was introduced into Entomologia Britannica on the authority of a fpecimen difcovered by ourfelves about ten years ago in the ifland of Anglefea, and from that period till very lately, when another example was taken in Devonfhire, this remained the only Britifh fpecimen known. It may hence be concluded that Cerambyx fanguineus is one of the moft uncommon as well as elegant fpecies of this genus found in Britain.

———

FIG. II. II.

CERAMBYX MINUTUS.

MINUTE CERAMBYX.

SPECIFIC CHARACTER

AND

SYNONYMS.

Rufous brown: antennæ as long as the body.

CERAMBYX MINUTUS: rufo-fufcous, antennis longitudine corporis. *Marfh Ent. Brit. T.* 1. *p.* 337. 21.

SAPERDA

SAPERDA MINUTA. *Fabr. Sp. Inf.* 1. 235. 2.

                        Mant. 1. 150. 39.

Callidium pygmæum. *Fabr. Ent. Syft.* 1. *b.* 323. 24.

A diminutive fpecies of very uncommon fcarcity: its appearance when magnified is particularly interefting.

F 2   PLATE

554

PLATE DLIV.

CARABUS CUPREUS.

COPPERY CARABUS.

COLEOPTERA.

GENERIC CHARACTER.

Antennæ filiform: feelers fix, the exterior joint obtufe and truncated: thorax obcordated, truncated behind and margined: wing-cafes margined: abdomen ovate.

SPECIFIC CHARACTER

AND

SYNONYMS.

Brafsy: antennæ red at the bafe.

CARABUS CUPREUS: æneus, antennis bafi rubris. *Linn. Fn. Suec.* 801.
Fab. Sp. Inf. 1. *p.* 308. *n.* 50.—*Mant. Inf.* 1. 201. 68.—*Ent. Syft.* 1. *a.* 153. 126.
Paykull Monogr. 71.
Fn. Fred. 21. 206.
Panz. Ent. Germ. 56. 60.
Illiger. Kugel. Kaf. Preus. 166. 31.
Marfh. Ent. Brit. T. 1. *p.* 439. 18.
Le Buprefte perroquet. *Geoffr.* 1. 161. 40.

The

The head, thorax, and wing-cafes are braffy brown, the under furface black with a violet glofs. This kind is diftinct from Carabus vulgaris, with which it might be confounded, efpecially in having the firft joint of the antennæ red, the whole of thefe organs being black in Carabus vulgaris.

PLATE

555

PLATE DLV.

PHALÆNA PHÆORRHŒA.

BROWN TAIL MOTH.

LEPIDOPTERA.

GENERIC CHARACTER.

Antennæ taper from the bafe: wings in general deflected when at reſt. Fly by night.

SPECIFIC CHARACTER

AND

SYNONYMS.

White: rays of the antennæ ferruginous: abdomen bearded and fuſcous at the end.

PHALÆNA CHRYSORRHŒA. *var. Linn.?*
BROWN-TAIL MOTH. *Curtis Hiſt. Brown-tail,* A. D. 1782.
PHALÆNA PHÆORRHŒA. *Marſh. Linn. Tranſ. V.* 1. *p.* 68.

———————

In the defcription of the 10th plate of this work we had occaſion to allude, in general terms, to an overſight committed by Linnæus in confounding the Yellow and Brown tail Moths under the fame name

as

as a fingle fpecies: the fubject reprefented in that plate is the Yellow-
tail, and the prefent feems requifite to complete the hiftory of thofe
two apparently ambiguous infects.

There is a diffimilarity, and that fo confiderable, between thofe two
infects, though at the firft view they may appear analogous, that, after
due comparifon, it muft excite furprize to learn they could have been
efteemed the fame by any competent Naturalift; yet they certainly
were, and not by Linnæus only; nor do they feem, even at this moment, to be very accurately defined as diftinct kinds by the generality
of continental writers, fome confidering them as varieties, and others
as the two fexes of an individual fpecies. Klemann is an exception
among thofe writers; he admits them to be diftinct on the authority
of Roefel, by whom both kinds were reared from the larvæ.

Befides thofe two moths, there is another more clofely allied to the
Yellow-tail than the Brown-tail, which has excited fome mifunderftanding; this is the infect called by Englifh collectors the " Spotted Yellow-
tail," as it differs from the former in having a large brown fhade along
the coftal margin beneath, and on the upper furface one or more obfcure dots. Fabricius, whofe opinion is countenanced by the authority of Villars and Schaeffer, defcribes it as a diftinct fpecies, under
the name of Auriflua, and this opinion is repeated in the work of
Gmelin: our Englifh collectors regard it, and not without probability,
as a fexual difference of the common Yellow-tail: we are perfuaded
it is no other than the male of that fpecies;—the male of the Brown-
tail Moth, we may further add, exhibits a fimilar appearance beneath.

The hiftory of the Brown-tail Moth is amply related in a little tract
publifhed about thirty years ago by the late Mr. W. Curtis, author
of the Flora Londinenfis. The occafion upon which that tract was
written is flightly mentioned in our defcription of the " Yellow-tail,"
and may now with propriety be repeated at greater length. The
period of time elapfed fince the appearance of Mr. Curtis's publication is not confiderable; yet, from the various viciffitudes to which
fuch a memorial of local events is neceffarily expofed, this interefting
pamphlet

pamphlet is now become fcarce : we fortunatcly poffefs it, and feeling perfuaded the information it conveys muft prove acceptable to the reader, fhall not neglect to introduce the moft material paffages for their perufal.

It will be within the recollection of many, that in the year 1782 the inhabitants of London and its vicinity were thrown into the utmoft confternation by the appearance of a phænomenon far from ufual in the northern regions of the earth; a hoft of infects, in numbers like the locufts of the deferts, were obferved at once to pervade the whole face of vegetation and defpoil the herbage in many places for miles of every trace of verdure:—thefe were no other than the larvæ of an infignificant Moth, the fubject of our prefent Plate.

The ravages committed by this infect were affuredly lefs confiderable than the vulgar were inclined to believe: true to their natural inftinct, fome particular vegetables were preferred to others, and thefe they devoured with impunity, while others were only partially attacked, as though eaten with reluctance in the general fcarcity of their natural food; and again, others being ftill lefs palatable, entirely efcaped their devastation. The afpect of vegetation was neverthelefs fuch as might juftly create alarm: plants, hedges, nay, whole plantations of fruit-trees, as well as trees of the foreft, fhared in the general havoc, prefenting their leaflefs branches in the midft of fummer, as though ftricken and deftroyed by the blafts of winter. An appearance fo extraordinary was calculated to create terror: it was naturally interpreted as a vifitation from heaven ordained to deftroy all the fources of vegetable life, to deprive men and cattle of their moft effential food, and finally leave them a prey to famine.—Such were the vulgar fears; but thanks to Providence, the deftroying powers of thefe creatures were reftricted by their inftincts; their attacks were principally directed againft the oak, the elm, the hawthorn, and fruit-trees: the fodder for the cattle and the harveft for mankind remained untouched. The appearance of fuch a hoft of little depredators feems, however, to have afforded a feafonable admonition, evincing to an unthinking

multitude how eafily the comforts, nay, even the very exiftence of man
may be affailed by a creature fo infignificant, had not the limits of its
ravages been prefcribed by Him " who wills and is obeyed;"—its in-
trufions certainly created alarm, but did little ferious injury.

This is no exaggerated picture of the public mind on the occafion
to which we refer; its alarm was fo powerful, and prevailed to fuch
an extent, that prayers were publicly offered up in the churches to
avert the calamity it was fuppofed they were intended to produce.
The webs containing the larvæ were collected in many places about
the metropolis by order of the parifh officers, who allowed a certain
price to the poor for gathering them, and fuperintended the burning
of them in large heaps with coal and faggots, a circumftance within our
own memory. At this precife period the tract by Mr. Curtis, as
above related, appeared. In this memoir the hiftory, manners, and
propenfities of this little creature were explained, and the information it afforded muft have undoubtedly contributed in an effential
manner to calm the terror before excited. Neither can we regard
its publication as being devoid of utility in another material refpect.
It muft furely have inclined the more reflecting part of the community, at leaft, to view the purfuits of the Entomologift, then confef-
fedly in a ftate of infancy in this country, with higher efteem than it
had been previoufly accuftomed to confider them.

" The attention of the public (fays Mr. Curtis) has of late been
ftrongly excited by the unufual appearance of infinite numbers of large
white webs, containing Caterpillars, confpicuous on almoft every hedge,
tree, and fhrub in the vicinity of the metropolis; refpecting which
advertifements, paragraphs, letters, &c. almoft without number, have
appeared in the feveral newfpapers, moft of which, though written
with a good intention, have tended greatly to alarm the minds of the
people, efpecially the weak and the timid. Some of thofe writers have
gone fo far as to affert, that they were an unufual prefage of the plague;
others, that their numbers were great enough to render the air pefti-
lential, and that they would mangle and deftroy every kind of vege-
                                                                    table,

table, and ftarve the cattle in the fields. From thefe alarming mif-representations, almoft every one ignorant of their hiftory has been under difmal apprehenfions concerning them; and even prayers have been offered up in fome churches to deliver us from the apprehended approaching calamity."—" Some idea may be formed (fays the fame author in a note on the above paffage) of their numbers from the following circumftances. In many parifhes about London fub-fcriptions have been opened, and the poor employed to cut off and collect the webs at one fhilling per bufhel, which had been burned under the infpection of the churchwardens, overfeers, or beadles of the parifh; at the firft onfet of this bufinefs, fourfcore bufhels, as I was moft credibly informed, were collected in one day in the parifh of Clapham."

One object in writing this tract was to fhew, that the infect was not new in this country, the fpecies being found every year, and in fome abundance, though not in plenty fufficient to excite the public attention. It was then known, as the author obferves, by thofe who collected infects as the caterpillar of the Brown-tail Moth. Nor is it peculiar to this country, being found in many parts of Europe. Albin, who publifhed in 1720, fays, the caterpillars lay themfelves up in webs all the winter, and as foon as the buds open they come forth and devour them in fuch a manner that whole trees, and fome-times hedges, for a great way together, are abfolutely bare. Geoffroy defcribes it as the moft common of all infects about Paris, where it is found on moft of the trees, which it often ftrips entirely of their foliage in the fpring. Our great naturalift Ray defcribes it likewife.

With refpect to the caterpillars of the Brown-tail Moth in the year 1782, and alfo in the year preceding, Mr. Curtis obferves, their numbers were uncommonly great and unufually extenfive, though he does not pretend to ftate the precife track in which they are found, having had no opportunity of obferving it, remarking only in this particular, that when infects are multiplied in this extraor-

c 2

dinary

dinary manner it is feldom that they extend through a whole country. " On the Kingfton road I traced them (fays this author) as far
as Putney Common, on the farther part of which, on the trees about
Coombe Wood and Richmond Park, a web was not to be feen.
I remarked, that they were extremely numerous at the diftance of
about eight miles on the Uxbridge road. On the great weftern
road they terminated about the Star and Garter leading to Kew;
from whence to Alton in Hampfhire not one was vifible; and I have
received undoubted information from other quarters, that the de-
ftruction they occafioned is by no means general."

Our remarks on the partiality fhewn by thefe infects for fome vegetables in preference to others will be eafily perceived from the
following ftatement: during the feafon mentioned (and in this they
are invariably conftant) they occurred on the *hawthorn* moft plentifully, *oak* the fame, *elm* very plentifully, *moft fruit-trees* the fame,
blackthorn plentifully, *rose-trees* the fame, and *bramble* the fame:
on the willow and poplar fcarce, and none were noticed on the
elder, the walnut, afh, fir, or herbaceous plants. Thus it appears,
that the principal injuries fuftained are in the orchard, the caterpillars deftroying the bloffoms as well as the leaves, and thereby
the fruit in embryo; the lofs of the leaves merely in many other
trees, fhould it happen in the fpring, being of fmall importance, as
thefe are reftored before the end of fummer.

Thefe caterpillars have happily many enemies; they are delec-
table food for moft birds, who eagerly devour them; they are alfo
victims to the Ichneumon fly, which deftroys them by myriads, and
it is fuppofed the abfence of the latter, from fome unknown caufe,
might have contributed, for one or two feafons, to their immenfe
increafe. The young caterpillars are hatched early in autumn. As
foon as they quit the egg they begin fpinning the web, and having
formed a fmall one, they proceed to feed on the foliage by eating,
like moft other larvæ, the upper furface and flefhy part of the leaf.
In thefe webs, which are progreffively increafed in fize as necef-

9   fity

fity requires, they live in focieties till they attain their laft fkin, when each fpins a feparate web or cocoon for itfelf: in this it paffes to the pupa form about the beginning of May, and after remaining a fhort time the Moth is produced *. There is more than one brood in a year, the fpecies being found in a winged in July and Auguft.

---

* It remains in the chryfalis about three weeks. *Curtis.*

PLATE

PLATE DLVI.

PHALÆNA VERTICALIS.

MOTHER-OF-PEARL MOTH.

LEPIDOPTERA.

GENERIC CHARACTER.

Antennæ taper from the bafe: wings in general deflected when at reft. Fly by night.

* *Section* PYRALIS.

SPECIFIC CHARACTER

AND

SYNONYMS.

Wings glabrous pale and fomewhat fafciated: beneath waved with fufcous.

PHALÆNA VERTICALIS: alis glabris pallidis fubfafciatis, fubtus fufco undatis. *Linn. Fn. Suec.* 1353.—*Gmel.* 2522. 835.

Phalæna Verticalis. *Fabr. Sp. Inf.* 2. *p.* 272. *n.* 180.—*Mant. Inf.* 2. *p.* 219. *n.* 285.

Geoffr. inf. Par. 2. *p.* 166. *n.* 112.

═════════════

Abundant in the month of July, when it appears in the winged ftate: the larva, which is of a delicate green colour, feeds on the common nettle: the pupa is diftinguifhed by having each of the three laft fegments armed or furnifhed with a tooth-like procefs.

PLATE

PLATE DLVII.

LEPTURA SANGUINOLENTA.

SANGUINEOUS LEPTURA.

COLEOPTERA.

GENERIC CHARACTER.

Antennæ fetaceous: feelers four, filiform: wing-cafes tapering towards the tip: thorax flender and round.

* Jaw with a fingle tooth, lip membranaceous and bifid. *Fabr.*

SPECIFIC CHARACTER

AND

SYNONYMS.

Black: wing-cafes fanguineous.

LEPTURA SANGUINOLENTA: nigra elytris fanguineis. *Linn. Syft. Nat.* 2. 638. 4.—*Fn. Suec.* 679.—*Fabr. Ent. Syft. T.* 1. *p.* 2. 341. 10.
Schaeff. Icon. tab. 39. fig. 9.

———

Rarely met with in Britain. Linnæus defcribes it as a native of Sweden, Schaeffer includes it among the infects found in the environs of Ratifbon, and we have feen it from Portugal. Fabricius fpeaks

VOL. XVI.   H   in

in general terms of the ſpecies as an inhabitant of Europe. It is found on flowers about the end of June or in the Month of July.

There is a variety of this inſect, having the wing-caſes margined with black; in the male, the wing-caſes are ſometimes tinged with teſtaceous, and at the tip with black.

PLATE

PLATE DLVIII.

NECYDALIS CÆRULEA.

BLUE NECYDALIS.

COLEOPTERA.

GENERIC CHARACTER.

Antennæ fetaceous or filiform: feelers four, filiform: wing-cafes lefs than the wings, and either narrower or fhorter than the abdomen: tail fimple.

SPECIFIC CHARACTER

AND

SYNONYMS.

Wing-cafes fubulate: body blue: pofterior thighs clavated and arched.

NECYDALIS CÆRULEA: elytris fubulatis cærulea femoribus pofticis clavatis arcuatis. *Fabr. Ent. Syft. T.* 1. *p.* 2. 354. 19.

NECYLALIS CÆRULEA. *Linn. Syft. Nat.* 2. 642. 4.

NECYDALIS CÆRULEA: cærulea, femoribus pofticis clavatis arcuatis. *Marfh. Ent. Brit. T.* 1. 359. 4.

Donov. Tour South Wales and Monmouthfhire. T. 1. *Glamorg.*

H 2   La

Telephorus cæruleus. *Degeer.* 4. 76. 8.
Cantharis nobilis. *Scop.*
La Cantharide verte à groffes cuiffes. *Geoff.* 1. 342. 3.
Ædemera cærulea. *Oliv.* 3. *p.* 50. 13. 16. *f.* 2. *f.* 16. *a. b.*

One of the moft fingular as well as beautiful fpecies of the infect tribe found in this country: it occurs on flowers of various kinds in in the middle of fummer: moft frequent on the golden cup, on the bramble and the dandelion, and perhaps we may fay generally, on umbelliferous flowers. In England it is rather a local infect; on the continent, and efpecially towards the South, it is obferved to be more abundant than in northern counties.

The moft material difference in the general appearance of the two fexes of this fpecies confifts in the ftructure of the pofterior legs; thefe in the female are fimple, but in the male are confiderably arched, while the thigh itfelf is fo remarkably large, in proportion to the reft of the legs and body, as to render its afpect particularly ftriking: the thigh is curved, and very globofe.

This difference in the ftructure of the thighs did not efcape the obfervation of Geoffroy, and after him of Fabricius; the latter of whom confiders the one with fimple legs merely as a variety of the other, " *Variat pedibus fimplicibus.*" Among the number of thofe who diffent from this idea we fhould, however, name that refpectable Naturalift, John Reinhold Forfter: he confidered it as a new fpecies, and defcribes it as fuch, under the appellation of Necydalis Ceramboides: " Elytris fubulatis, viridi-ænea pedibus fimplicibus," in his tract of *One Hundred* new Species of Infects. Fabricius, who wrote after the time of Forfter, refers to this among his fynonyms, and feems therefore convinced it can be no other than a variety of Cærulea.

In

In Entomologia Britannica we find the opinion of Forſter preferred to that of Fabricius, the ſuppoſed ſpecies being included under the name of Ceramboides, as in " Novæ Species Inſectorum." Mr. Marſham, neverthelefs, with that degree of caution which ſo eminently diſtinguiſhes his valuable work, expreſſes a doubt whether it ought to be conſidered as a diſtinct ſpecies or a ſexual difference, " *An ſpecies diſtincta? An ſexús differentia ?*"

The ſmaller figures denote the natural ſize; the variation that prevails in colour from a blue to bright braſſy green is ſhewn in the larger figures.

PLATE

PLATE DLIX.

MUSCA EPHIPPIUM.

RUFOUS-THORAX MUSCA.

DIPTERA.

GENERIC CHARACTER.

Mouth with a fhort exferted flefhy probofcis, and two equal lips: fucker furnifhed with briftles: feelers two, very fhort, or fometimes none: antennæ generally fhort.

* Stratiomys. *Fabr.*

SPECIFIC CHARACTER

AND

SYNONYMS.

Scutel furnifhed with two teeth: thorax rufous, fpinous each fide.

MUSCA EPHIPPIUM: fcutello bidentato, thorace rufo utrinque fpinofo. *Fabr. Sp. Inf.* 2. *p.* 417. *n.* 2.—*Mant. Inf.* 2. 330. *n.* 4.—*Ent. Syft. T.* 4. 264. *n.* 6. *Geoffr. Inf.* 2. 480. 3.

MUSCA INDA. *Schranck. Inf. Auftral. p.* 438. *n.* 891.

———————

Taken in Coombe Wood, on the 4th of June, 1812, by Geo. Milne, Efq. F. L. S.

Mufca ephippium is confidered as a fcarce infect in this country. It is known to have been taken occafionally by the old collectors about

thirty

thirty years go, or rather more, in the woods about Highgate, which, in confequence of the recent improvements in that vicinity of the me-troplis, are now demolifhed. It has occurred likewife in the woods of Kent. Our good friend, John Swainfon, Efq. of Liverpool, met with three fpecimens at the fame time fticking againft the trunks of trees; this happened about twenty years ago, and was then efteemed a rare circumftance.

There is no other fpecies of Mufca that can be confounded with this; the characteriftic rufous colour of the thorax forms fuch a ftriking contraft with the black of the body and the dufkinefs of the wings as to render this impoffible: neither do we recollect, among the innumerable tribes of exotic fpecies, any infect fo ftrikingly fingular, with the exception of fome of the Mutilla genus, the very different ftructure of which can never allow them to be miftaken for Mufca, even by the moft cafual obferver. When clofely examined, the thorax is perceptibly covered with fhort hairs; the body is quite fmooth and black, the legs black, and poifers pale yellowifh. The whole furface beneath is black.

P L A T E

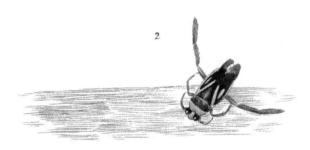

PLATE DLX.

FIG. I.

NOTONECTA MACULATA.

SPOTTED BOAT-FLY.

HEMIPTERA.

Snout inflected: antennæ fhorter than the thorax: wings four folded crofs-wife, coriaceous on the upper half: pofterior legs hairy, and formed for fwimming.

SPECIFIC CHARACTER

AND

SYNONYMS.

Wing-cafes fufcous, springled with ferruginous fpecks, the tip bifid.

NOTONECTA MACULATA: elytris fufcis ferrugineo irroratis apice bifidis. *Fabr. Ent. Syft. T. 4. p. 58. 3.*

Found in the waters about London, and in Devonfhire: Fabricius defcribes it as an inhabitant of waters about Paris on the authority of Bofc.

VOL. XVI.   I   FIG.

FIG. II.

NOTONECTA FURCATA.

FURCATE BOAT-FLY.

SPECIFIC CHARACTER

AND

SYNONYMS.

Wings fuscous with two testaceous spots at the base, the tip bifid.

NOTONECTA FURCATA: elytris fuscis: maculis duabus baseos testaceis, apice bifidis. *Fabr. Ent. Syst. T. 4. p.* 58. 2.

―――――――――

Less abundant than Notonecta glauca, to which it is nearly allied in size and appearance: the two oblong testaceous spots at the base of the wing cases sufficiently distinguish it from N. glauca. Both have the same haunts and manners of life. N. furcata has been found, according to Bosc, in the waters about Paris.

P L A T E

561

PLATE DLXI.

SCARABÆUS VACCA.

HORNED, or COW-HEADED BEETLE.

COLEOPTERA.

GENERIC CHARACTER.

Antenuæ clavated, the club lamellate: feelers four: anterior fhanks generally toothed.

SPECIFIC CHARACTER

AND

SYNONYMS.

Exfcutellate: thorax unarmed, acuminate: head armed with two erect fpines.

SCARABÆUS VACCA: exfcutellatus thorace mutico acuminato, occipite fpina erecta gemina. *Linn. Syft. Nat.* 2. 547. 23.

SCARABÆUS VACCA. *Fabr. Syft. Ent.* 26. 101. *Sp. Inf.* 1. 28. 126.—*Mant.* 1. 15. 143.—*Ent. Syft.* 1. a. 55. 179.

SCARABÆUS VACCA. *Gmel. Syft.* 1543. 25.

Marfh. Ent. Brit. T. 1. *p.* 34. 61.

Copris. *Geoff.* 1. 90. 5.

Copris confpurcatus. *Fourc.* 14. 5.

Onthophagus Vacca. *Latr. Gen. Cruft. et Inf.* T. 2. *p.* 87.

1 2   An

An infect of very remarkable and interefting figure, efpecially in the front view, the horns on the head refembling thofe of the cow. The male only is furnifhed with horns, the head of the female, as in many other fpecies of this tribe, having only a flight protuberance inftead. It is fcarce in England. Has been ufually taken in cow-dung.

The fmaller figure reprefents the natural fize.

P L A T E

PLATE DLXII.

PHALÆNA SIGMA.

DOUBLE SQUARE SPOT.

LEPIDOPTERA.

GENERIC CHARACTER.

Antennæ taper from the bafe: wings in general deflected when at rest: fly by night.

SPECIFIC CHARACTER

AND

SYNONYMS.

Wings purplifh brown with pale bands, and a double fquare black fpot in the middle.

NOCTUA SIGMA: criftata, alis deflexis, fuperioribus mofchatinis כ fufco nigro infcriptis. *Klem.* 2. *p.* 10. 25.

NOCTUA SIGNUM: criftata alis maculis tribus fufcis: cofta bafeos cinerafcente, thorace fufco antice brunneo. *Fabr. Mant. Inf. T.* 2. *p.* 154. *n.* 141.

NOCTUA SIGMA. *Knoch Beitr.* 3. *p.* 94. 10.

NOCTUA ATROSIGNATA. *Wiener Verz. p.* 78.

———

Found in the larva ftate in May and June, appears on the wing in Auguft. The fpecies is well diftinguifhed by the character-like fufcous mark in the middle of the anterior wings, a kind of double mark, formed by the junction of two fubquadrangular fpots that unite together at their moft contiguous angle, being a little oblique from each other. Some compare this mark to the Hebrew character כ, but it does not well agree with this; the fimple appellation of double fquare fpot is much more applicable.

PLATE

PLATE DLXIII.

STAPHYLINUS HYBRIDUS.

HYBRID ROVE BEETLE.

COLEOPTERA.

GENERIC CHARACTER.

Antennæ moniliform: feelers four: wing-cafes half as long as the body: wings folded up under the wing-cafes: tail fimple, and fur-nifhed with two exfertile veficles.

SPECIFIC CHARACTER.

AND

SYNONYMS.

Pubefcent, golden-fulvous, or greyifh clouded with blackifh: abdomen black at the tip: thighs annulated with yellow.

STAPHYLINUS HYBRIDUS: fulvo-aureo-pubefcens nigro nebulofus, abdomine apice nigro, femoribus annulo flavo. *Marfh. Ent. Brit. T.* 1. *p.* 500. 9.

▬▬▬▬

A new fpecies difcovered by the Rev. Mr. Kirby in the middle of October, 1799. Since that period it has been found by Dr. Leach and other collectors. It appears to be not very uncommon in fome places; and ufually occurs in the dung of animals.

PLATE

PLATE DLXIV.

TABANUS NIGER.

BLACK TABANUS.

DIPTERA.

GENERIC CHARACTER.

Mouth with a ftraight exferted membranaceous probofcis, terminated in two equal lips: fucker projecting, exferted, and placed in a groove on the back of the probofcis, with a fingle-valved fheath and five briftles: feelers two, equal, clavate, and ending in a point: antennæ fhort, approximate, cylindrical, with an elevated tooth at the bafe.

SPECIFIC CHARACTER.

TABANUS NIGER. Black: eyes filky green, with the anterior margin and three bands of purple.

A new and very curious fpecies, lately difcovered in Hampfhire by G. Montagu, Efq. The figures in the annexed plate will fhew the upper and lower furfaces as they appear when magnified. The fmaller figure exemplifies the natural fize.

VOL. XVI.   K   PLATE

PLATE DLXV.

CARABUS DIMIDIATUS.

KUGELANNIAN CARABUS.

COLEOPTERA.

GENERIC CHARACTER.

Antennæ filiform: feelers fix, the exterior joint obtufe and truncated: thorax obcordated, truncated behind, and margined: wing-
cafes margined: abdomen ovate.

SPECIFIC CHARACTER

AND

SYNONYMS.

Braffy green: head, thorax, and outer margin of the wing-cafes
generally purple: wing-cafes fomewhat convex ftriæ: legs black.

CARABUS DIMIDIATUS: æneo-virens, elytris ftriatis: interftitiis con-
vexiufculis, pedibus nigris. *Marfh. Ent. Brit.*
T. 1. *p.* 445. *Sp.* 35.
Oliv. Inf. 111. 35. 72. 94. *t.* 11. *f.* 121.
CARABUS KUGELANNII. *Panz. Faun. Germ.* 89. *t.* 8.
Illiger. Kugel. Kaf. Preuf. 166. 30.

A very fcarce infect, and one of the moft beautiful of the Britifh
Carabi. Its fize is moderate, or rather fmall, fomewhat exceeding
the length of half an inch, the colours on the fuperior furface vivid,
beneath black with a faint glofs of violet. The antennæ are black

K 2

with

with the bafe rufous, the legs entirely black. The head and thorax fmooth, polifhed, and gloffy, and the wing-cafes ftriated; the interftices deep, and marked with impreffed dots in a fingle feries.

The colours in this fpecies are not conftant, the head and thorax in fome inftances being greenifh and deftitute of the fine purple tinge, fo confpicuous in the fpecimen we have reprefented.

PLATE

PLATE DLXVI.

CHRYSOMELA ATRICILLA.

BLACK-HEADED CHRYSOMELA.

COLEOPTERA.

GENERIC CHARACTER.

Antennæ moniliform: feelers fix, growing larger towards the end: thorax marginate: wing-cafes immarginate: body oval.

SPECIFIC CHARACTER

AND

SYNONYMS.

Head black: thorax, wing-cafes, and fhanks teftaceous.

CHRYSOMELA ATRICILLA: nigra, thorace elytris tibiifque teftaceis,
futura nigra. *Marfh. Ent. Brit. T.* 1. 200. 74.
—*Linn. Syft. Nat.* 594. 55.—*Gmel.* 1693. 55.
ALTICA ATRICILLA. *Fab. Syft. Ent.* 115. 17.
GALLERUCA ATRICILLA. *Fab. Ent. Syft.* 1. b. 31. 89.

A fmall fpecies: the breaft, abdomen, and pofterior thighs are black.

PLATE

PLATE DLXVII.

PAPILIO MALVÆ.

MALLOW, OR GRIZZLED SKIPPER BUTTERFLY.

LEPIDOPTERA.

GENERIC CHARACTER.

Antennæ clavated at the tip: wings erect when at rest. Fly by day.

SPECIFIC CHARACTER

AND

SYNONYMS.

Wings indented divaricate, brown waved with cinereous: anterior pair with hyaline dots: posterior with white dots beneath.

PAPILIO MALVÆ: alis dentatis divaricatis fuscis cinereo undatis: anticis punctis fenestratis, posticis subtus punctis albis. *Linn. Syst. Nat.* 2. 795. 267.—*Fn. Su.* 1081.

HESPERIA MALVÆ. *Fabr. Ent. Syst.* 3. 350. 333.—*Syst. Ent.* 535. 396. *Sp. Ins.* 137. 638.

———————

The larva of this Butterfly feeds on the mallow: the colour is greyish or yellowish, with the head black, and a black collar marked with four sulphur-coloured spots. The pupa is somewhat gibbous and blueish.

This

This infect is common in many parts of Britain in the fly ftate; the larva, though known, by no means common. The Butterfly appears on the wing in May.

Some collectors admit two or more varieties of the Grizzled Skipper Butterfly, while others confider them as fo many diftinct fpecies: the male alfo differs a little from the female in being fome-what fmaller.

PLATE

PLATE DLXVIII.

PHALÆNA LUBRICIPEDA.

SPOTTED BUFF MOTH.

LEPIDOPTERA.

GENERIC CHARACTER.

Antennæ taper from the bafe: wings in general deflected when at rest; fly by night.

SPECIFIC CHARACTER

AND

SYNONYMS.

Wings yellowifh, with black dots generally in an oblique tranfverfe row.

PHALÆNA LUBRICIPEDA. *Marfh. Linn. Tranf. T.* 1. *p.* 71. *tab.* 1. *fig.* 2.
BOMBYX LUBRICIPEDA. *Linn. Fn. Sv.* 1138. *mas.*
Fabr. Syft. Ent. 576. 68.

———————————

The larva of this kind is hairy and brownifh, with a lateral white ftripe: it feeds on herbaceous plants, and is found in Auguft. The fly appears in June.

VOL. XVI.   L   PLATE

1

1

2

2

PLATE DLXIX.

FIG. I.

CHRYSOMELA NEMORUM.

COLEOPTERA.

GENERIC CHARACTER.

Antennæ moniliform: feelers fix, growing larger towards the end: thorax marginate: body in general oval.

SPECIFIC CHARACTER

AND

SYNONYMS.

Black: ftripe down the middle of the wing-cafes, and the legs yellow.

CHRYSOMELA NEMORUM: atra, elytris linea flava, pedibus flavis. *Linn. Syft. Nat.* 595. 62.—*Fn. Suec.* 543.— *Gmel.* 1695. 62.—*Marfh. Ent. Brit. T.* 1. 197. 65.
ALTICA NEMORUM. *Fabr. Syft. Ent.* 115. 20.—*Panz. Ent. Germ.* 181. 27.
GALLERUCA NEMORUM. *Fabr. Ent. Syft.* 1. *b.* 31. 104.
L'Altife à bandes jaunes. *Geoff.* 1. 247. 9.

━━━━━━━━━

This minute fpecies is extremely common in fome fituations. The body is of an oblong fhape, and the legs formed for leaping.

L 2   F I G.

FIG. II.

CHRYSOMELA MODEERI.

MODEER'S CHRYSOMELA.

SPECIFIC CHARACTER

AND

SYNONYMS.

Braffy black: wing-cafes at the tip yellow: four anterior legs, with the fhanks of the pofterior ones yellow.

CHRYSOMELA MODEERI: ænea nitida, elytris apice flavis pedibus anterioribus tibiifque pofticis luteis. *Linn. Syft. Nat.* 594. 57.—*Fn. Su.* 539.—*Marfh. Ent. Brit. T.* 1. *p.* 194. 56.

Altica Modeeri. *Panz. Ent. Germ.* 177. 9.

Galleruca Modeeri. *Fabr. Ent. Syft.* 1. *b.* 30. 85.

───────────

Size of the former.

PLATE

PLATE DLXX.

CURCULIO DIDYMUS.

DOUBLE-SPOTTED CURCULIO, or WEEVIL-BEETLE.

COLEOPTERA.

GENERIC CHARACTER.

Antennæ clavated, and feated on the fnout, which is horny and prominent: feelers four, filiform.

SPECIFIC CHARACTER.

CURCULIO DIDYMUS. Cinereous, thorax with a longitudinal dorfal carination: wing-cafes with raifed ftriæ, and a feries of double or confluent fpots between each.

─────────

A new fpecies, allied in appearance to C. carinatus of Fabricius. It was firft difcovered by Mr. Rope, who communicated the fpecimen to the author of Entomologia Britannica, T. Marfham, Efq. and fince that period the fame fpecies has been taken in Coombe Wood. Dr. Leach has a fpecimen, found in the place laft mentioned.

When magnified the appearance of this infe&t is very remarkable. The general colour is cinereous inclining to blackifh: down the middle of the thorax is a diftin&t longitudinal carinated or raifed line; the wing-cafes are marked with a few raifed longitudinal lines, and the interftices impreffed with a feries of double pun&tures or dots, or rather two feries of dots placed nearly parallel, and every pair fo clofely approximating down the middle of the interftices as to appear like a double dot.

PLATE

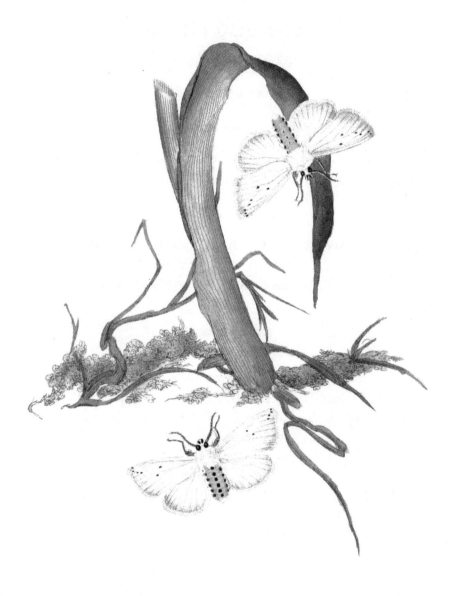

PLATE DLXXI.

PHALÆNA PAPYRATIA.

WATER ERMINE-MOTH.

LEPIDOPTERA.

* *Bombyx.*

GENERIC CHARACTER.

Antennæ taper from the bafe: wings in general deflected when at
at reft.   Fly by night.

SPECIFIC CHARACTER

AND

SYNONYMS.

Wings fnowy white with black dots at the tip: abdomen with five
rows of black dots.

BOMBYX PAPYRATIA.   *Marfh. Linn. Tranf.* 1. *p.* 72. *tab.* 1.
fig. 4.

Refembles the large or common Ermine Moth, and feems to have
been very frequently confounded with that fpecies till its fpecifical
diftinction was pointed out by our worthy friend Thomas Marfham,
Efq. in a memoir printed in the firft volume of the Tranfactions of
the Linnæan Society. It differs principally in having black dots at
the tip of the wings only, except one or two reaching in a line towards

---

* B. Menthraftri of Fabricius.

the

the bafe: the abdomen fulvous, and the tip white. In P. Erminea the black dots on the wings are more numerous.—We muft, however, add, that, in fome inftances, the wings of Bombyx papyratla occurs with fcarcely any black dots, The female has alfo, in general, fewer fpots than the male.

This fpecies in the larva ftate feeds on aquatic plants, and, as the trivial name implies, is ufually found in watery places in the winged state. The larva is fufcous and hairy; pupa black.

PLATE

572

PLATE DLXXII.

FIG. I.

COCCINELLA 5-PUNCTATA.

FIVE-DOT RED COCCINELLA, or COW-LADY.

COLEOPTERA.

GENERIC CHARACTER.

Antennæ clavated, the club folid: anterior feelers hatchet-fhaped, pofterior filiform: thorax and wing-cafes margined: body hemifpherical: abdomen flat.

SPECIFIC CHARACTER

AND

SYNONYMS.

Wing-cafes fanguineous with five black dots.

COCCINELLA 5-PUNCTATA: coleoptris fanguineis: punctis nigris quinque. *Linn. Syft. Nat.* 580. 11.—*Fn Suec.* 474.—*Fabr. Syft. Ent.* 80. 11.—*Sp. Inf.* 1. 96. 17.—*Mant.* 1. 56. 31.—*Ent. Syft.* 1. a 278. 36. *Marfh. Ent. Brit. T.* 1. 151. 5.

━━━━━━━━

The wing-cafes in this fpecies are red, with two black dots on each, and one common black dot at the bafe. The thorax is black, with the anterior angle white.

VOL. XVI.   **M**   **FIG.**

FIG. II.

COCCINELLA 11-PUNCTATA.

11-DOT COCCINELLA, or COW-LADY.

SPECIFIC CHARACTER

AND

SYNONYMS.

Red, with eleven black dots.

COCCINELLA 11-PUNCTATA: coleoptris rubris: punĉtis nigris un-
decim. *Linn. Syſt. Nat.* 581. 15.—*Fn. Suec.*
480.—*Gmel.* 1561. 18.—*Fab. Syſt. Ent.* 82.
19.—*Sp. Inſ.* 1. 98. 31.—*Mant.* 1. 57. 46.—
Ent. Syſt. 1. *a* 277. 53.—*Marſh. Ent. Brit. T.*
1. *p.* 155. 16.

Considered by Paykull as a variety of Coccinella collaris.

PLATE

PLATE DLXXIII.

FIG. I. I.

STAPHYLINUS ANGUSTATUS.

NARROW ROVE-BEETLE.

COLEOPTERA.

GENERIC CHARACTER.

Antennæ moniliform: feelers four: wing-cafes half as long as the body: wings folded up under the cafes: tail not armed with a forceps, furnifhed with two exfertile veficles.

SPECIFIC CHARACTER

AND

SYNONYMS.

Filiform, black: tip of the wing-cafes and legs teftaceous.

STAPHYLINUS ANGUSTATUS: filiformis ater elytris apice pedibufque teftaceis. *Paykull. Monogr.* 36. 27.—*Fabr.
Ent. Syft. T.* 1. *p.* 2. *p.* 528. *Panz. Ent. Germ.*
356. 31.
STAPHYLINUS ANGUSTATUS. *Marfh. Ent. Brit. T.* 527. 83.

━━━━━━━━━

A fmall fpecies.

M 2   FIG.

F I G. II. II.

STAPHYLINUS BIGUTTATUS

BIGUTTATE ROVE-BEETLE.

SPECIFIC CHARACTER

A N D

SYNONYMS.

Black : wing-cafes with a yellow dot.

STAPHYLINUS BIGUTTATUS: niger, elytris puncto flavo. *Linn.*
Syſt. Nat. 685. 15.—*Fn. Su.* 851.—*Gmel.* 2029.
15.—*Fab. Ent. Syſt.* 1. *b.* 527. 36.—*Sp. Inf.* 1.
336. 13.—*Marſh. Ent. Brit. T.* 1. *p.* 526. 81.
STAPHYLINUS JUNO *var β. Paykull Monogr.* 25.
LE STAPHYLIN JUNON. *Geoffr.* 1. 371. 24.

Twice the fize of the former.

STAPHYLINUS ELONGATUS.

ELONGATED ROVE-BEETLE.

SPECIFIC CHARACTER

A N D

SYNONYMS.

Black : wing-cafes behind, with the legs, and antennæ ferruginous.

STAPHYLINUS

STAPHYLINUS ELONGATUS: niger, elytris poftice pedibus anten-
nifque ferrugineis.　*Linn. Syft. Nat.* 685. 14.—
Marfh. Ent. Brit. T. 1. *p.* 515. 52.
Paederus elongatus.　*Fabr. Syft. Ent.* 268. 2.

——————

Oblong, and glabrous; found in dung.

PLATE

574

PLATE DLXXIV.

MUSCA TENAX.

DIPTERA.

GENERIC CHARACTER.

Mouth with a foft exferted flefhy probofcis with two equal lips: fucker furnifhed with briftles: feelers two, very fhort or none: antennæ ufually fhort.

* *Syrphus.*

SPECIFIC CHARACTER

AND

SYNONYMS.

Downy: thorax grey: abdomen brown: hind-fhanks comprefled and gibbous.

MUSCA TENAX. *Linn. Syft. Nat.* 2. 984. 32.—*Fn. Su.* 1799.
SYRPHUS TENAX: antennis fetariis tomenofus thorace grifeo, abdomine fufco, tibiis pofticis comprefio gibbis. *Fabr. Ent. Syft. T.* 4. 288. 36.

Found in dung and putrid fubftances.

PLATE

PLATE DLXXV.

HISPA MUTICA.

HAIRY UNARMED HISPA.

COLEOPTERA.

GENERIC CHARACTER.

Antennæ cylindrical, approximated at the bafe, and feated between the eyes: feelers fuciform: thorax and wing-cafes ufually fpinous or toothed at the tip.

SPECIFIC CHARACTER

AND

SYNONYMS.

Unarmed, black: antennæ hairy: wing-cafes ftriated.

Hispa mutica; inermis, nigra, antennis pilofis, elytris ftriatis.— *Linn. Syft. Nat.* 604. 4.—*Vill.* 1. 170. 3.— *Gmel.* 1732. 4.—*Fabr. Syft. Ent.* 71. 6.—*Sp. Inf.* 1. 83. 9.—*Mant.* 1. 477.—*Marfh. Ent. Brit.* T. 1. p. 232.
Ptilinus muticus. *Fab. Ent. Syft.* IV. *App.* 443.
Dermeftes clavicornis. *Linn. Fn. Su.* 413.
Tenebrio hirticornis. *Degeer.* v. 47. t. 3. f. 1.

―――――――

A minute fpecies; in its manners faid to refemble the Dermeftes tribe, being like that infect found among fur, leather, clothes, &c. to which it proves injurious. The head is exferted, the thorax angulated; antennæ filiform, and thickeft in the middle.

VOL XVI.   N   PLATE

PLATE DLXXVI.

PHALÆNA FASCELINA.

DARK TUSSOCK MOTH.

LEPIDOPTERA.

GENERIC CHARACTER.

Antennæ taper from the bafe : wings in general deflected when at rest : fly by night.

SPECIFIC CHARACTER

AND

SYNONYMS.

Wings deflected, cinereous, fprinkled with black points, and tra-verfed by two flexuous fulvous ftreaks.

PHALÆNA FASCELINA : alis deflexis cinereis : atomis ftrigifque duabus repandis. *Linn. Syft. Nat.* 2. 825. 55. *Fn. Su.* 1119.—*Fabr. Ent. Syft. T. 3. p. 1. p.* 439. 98.

———

The larva of this Moth is hairy and tufted, the pupa folliculate. The larva is found (chiefly on the oak) in the month of May ; the Moth appears in July. A local fpecies, common in fome parts of the country, but not frequent near London.

N 2   LINNÆAN

LINNÆAN INDEX

TO

VOL. XVI.

———

COLEOPTERA.

INDEX.

———

HEMIPTERA.

———

LEPIDOPTERA.

———

NEUROPTERA.

DIPTERA

INDEX.

DIPTERA.

ALPHABETICAL

ALPHABETICAL INDEX

TO

VOL. XVI.

VOL. XVI.   O   montana,

INDEX.

ALPHA-

ALPHABETICAL TABLE

OF

CONTENTS.

VOL. XI. TO VOL. XVI.

o 2   cephalotes

CONTENTS.

CONTENTS.

CONTENTS

variegatis,

CONTENTS.

Printed by Law and Gilbert, St. John's-Square, London.